SUCCESSFUL SURVEYS

RESEARCH METHODS AND PRACTICE

Fourth Edition

NEIL GUPPY
GEORGE GRAY

THOMSON NELSON

Australia Canada Mexico Singapore Spain United Kingdom United States

THOMSON
NELSON

Successful Surveys: Research Methods and Practice
Fourth Edition

by Neil Guppy and George Gray

Associate Vice President Editorial Director:
Evelyn Veitch

Editor-in-Chief, Higher Education:
Anne Williams

Executive Editor:
Cara Yarzab

Director of Marketing:
Kelly Smyth

Developmental Editor:
Glen Herbert

Content Production Manager:
Tannys Williams

Production Service:
ICC Macmillan Inc.

Copy Editor:
Elizabeth Phinney

Proofreader:
Joan Templeton

Indexer:
Julie Nemer

Manufacturing Coordinator:
Loretta Lee

Design Director:
Ken Phipps

Interior Design:
Katherine Strain

Cover Design:
Johanna Liburd

Cover Image:
Daly & Newton/Getty Images

Compositor:
ICC Macmillan Inc.

Printer:
Thomson West

Printed and bound in the United States
2 3 4 10 09 08 07

For more information contact Nelson, 1120 Birchmount Road, Toronto, Ontario. M1K 5G4. Or you can visit our Internet site at http://www.nelson.com

Library and Archives Canada Cataloguing in Publication

Guppy, L. Neil, 1949–
Successful surveys: research methods and practice / Neil Guppy, George Gray.—4th ed.

ISBN-13: 978-0-17-610294-4
ISBN-10: 0-17-610294-9

1. Social surveys—Textbooks. 2. Social sciences—Research—Methodology. I. Gray, George A. II. Title.

HN29.G73 2007 300.72'3
C2006-906734-1

Preface

Teaching well is a difficult process, requiring enormous skill and dedication. Writing a book to assist in this process is, we have found, a daunting task. Both of us thought we knew a lot when we first started teaching. However, we were quickly forced to recognize our arrogance and rethink our earlier assumptions. The lesson was not particularly well learned, however, because we assumed that writing a book on survey research would be a simple matter—we had piles of lecture notes and hours of classroom experience. In both cases, teaching a course and writing a book, we had to learn by doing.

People learn by doing, and that principle, learning by experience, is a fundamental part of this book. What we do in the pages that follow is take readers through the survey process. The emphasis is on doing rather than on discussing the issues involved in survey research methodology.

The book is intended for college and university students who are taking a first course in research methods or survey design. So far as possible, the language is kept simple to increase readability. Readers may associate simplicity in language with simplicity in conducting a survey, but we try hard to prevent that misconception. The book is also intended for students from a variety of social science disciplines—commerce, economics, education, geography, health care, planning, political science, recreation, and sociology. All of these disciplines rely on surveys as a major source of data. The examples we use reflect the diversity of these disciplines.

The main focus of the book is on developing two sets of general skills for the student. The first set of skills has to do with formulating a good research question. The art of framing a researchable question is something that sounds easy but is extremely difficult in practice. The second set of skills involves the technical capacity to undertake high-quality research using survey methods. These latter skills include some obvious abilities such as learning how to phrase questions, and some less obvious skills such as assigning understandable codes to replies so that computers can properly analyze response patterns.

The emphasis is on *doing* survey research. We pay careful attention to presenting the necessary tools for doing good work, stressing application over pontification. A set of exercises, each building on the previous one, guides students through the process of survey research. Students start by defining a research question and end with outlining a research report.

Unlike many books on survey research, we try hard to stress non-academic applications. The majority of students neither become academics nor get involved in academic research (except possibly as subjects). They may, however, undertake survey research activity for themselves or for other people (e.g., in marketing firms or public-sector research departments). Many of our examples reflect this non-academic focus, and many of our ethical issues are concerned with the relationship between the researcher and the client.

Instructors who use the book in their courses may find our chapter ordering a little unconventional. For example, many books defer discussions of ethical issues until the end. We feel this is a serious mistake, since doing a survey might be a very bad decision based on ethical grounds. Ethical issues should be decided very early when considering a research project. We also have students formulate a research question before presenting the more abstract issues of causality, spuriousness, scope conditions, and the like. This decision is based on the idea that people will learn more effectively if they have a concrete example to which the lessons will apply.

Students should be aware that surveys do not follow a linear, developmental model. In any book, chapter follows chapter in a linear progression. Unfortunately (for book writing), survey research does not follow this linear logic. Book chapters give the appearance of independent blocks of knowledge, but survey design issues are intertwined. In doing a survey, you will find that decisions do not occur mechanically, one after another, as book chapters do. What questions you ask relate to how you ask questions, and both of these issues involve ethical principles. Survey refinements in one area have consequences for all other decisions. So while the book has a linear form, do not be deceived—good survey research involves a constant monitoring and basic clarification of both purpose and design.

We do not devote much attention to the analysis of data. Our reasons for this are numerous, but include the fact that many researchers in the social sciences have paid far too much attention to the sophisticated analysis of data, and far too little attention to improving the quality of data collection. Complex statistical analyses have, in our judgment, been oversold. As well, a growing number of colleges and universities separate the teaching of statistics and research methodology. There are many good statistical texts and a growing array of computer software packages that can be used with this book if the need arises.

We have made several changes from the previous editions. First, our discussion of the Internet is further expanded since the Web in particular has had profound implications for survey research. We have integrated material that reflects these exciting changes throughout the book. Second, we have added more examples, updated references, and appended illustrations of actual survey questions. Third, and following the suggestions of instructors who use this book, we have amalgamated some chapters to provide a better flow. Finally, in every chapter we have reworked old material or introduced new ideas and new research in order to help students better understand the complexities of a successful survey.

ACKNOWLEDGMENTS

We owe a great deal to the many students at UBC, from a variety of fields, who have taken courses in survey design with us. Their insights have been rewarding and educational. We are indebted also to our colleagues Martha Foschi, Patricia Marchak, Gerry Veenstra, and especially Martin Meissner, not so much for anything they did in helping us get this book together, but more for their challenging approaches to all manner of social science issues. Arguments and discussions with them have influenced us far more than even we probably realize. The ways in which we were taught research methods also influence these pages, and we must thank Alfred Hunter, Powhatan Wooldridge, and W.S. Robinson. We also gratefully acknowledge some early work by Sophia Lum that got this project started and some wonderful assistance with earlier editions by Augustine Park and Robin Hawkshaw. We owe a debt also to Mary Ann Clarke for her wonderful illustrations that help enliven the book while assisting in the learning process (any similarities between her "Sam Snerd" and her survey methods instructor are purely coincidental!). This book was improved upon significantly after the careful commentary we received, with thanks, from Laura Johnson, Patricia Pakvis, Katherine Watson, and Beth Mawhiney. Finally, our thanks to the team at Thomson Nelson (and previously Harcourt Brace & Company), who have contributed to the development of this and previous editions, including Glen Herbert, Cara Yarzab, Sheila Malloch, Lorraine Doherty, Heather McWhinney, Dan Brooks, Celène Adams, James Bosma, Stacey Roderick, Sarah Robertson, Wendy Thomas, Michael Bolton, Emily Ferguson, Brad Lambertus, Eliza Marciniak, and Elizabeth Phinney.

Statistics Canada information is used with the permission of the Minister of Industry, as minister responsible for Statistics Canada. Information on the availability of the wide range of data from Statistics Canada can be obtained from Statistics Canada's regional offices, its website at http://www.statscan.ca, and its toll-free access number: 1-800-263-1136.

Contents

Chapter 1

INTRODUCTION TO SURVEY RESEARCH

Have you ever participated in a survey? Chances are you have. Many people have answered the telephone and heard a voice asking for help in answering some questions, questions perhaps about voting preferences or TV viewing habits. Others may have been stopped in a shopping mall and asked questions about a wide range of issues, from local politics to preferences in colas. Still others will have responded to surveys on the Web. Also, someone in every Canadian household completes a census questionnaire—a government-sponsored "survey" of all Canadians.

Surveys involve collecting information by asking people specific questions. Identical questions are asked of everyone participating in a survey, and responses are categorized using a common coding scheme. Since everyone is asked the same questions, researchers are able to compare in a systematic way the responses people give. This type of systematic information gathering is valuable to decision-makers. For example, when political leaders are concerned about the timing of elections, surveying voter opinion is helpful. If someone wants to know what types of people are likely to engage in volunteer work, information gathered using a survey can help. The wide range of topics on which systematic survey information can be collected makes surveys a useful tool. Surveys have, therefore, become a fact of life.

THE EARLIEST SURVEYS

It is no coincidence that the words *state* and *statistics* share similar roots. Some of the earliest surveys were done by heads of state in Rome and Egypt. They wanted data on two things: how many warriors they had for military campaigns and how much people could afford to pay in taxes! Population counts, obtained using survey methods, remain valuable to the state.

SURVEYS IN HISTORY

By reviewing briefly the history of surveys, we can highlight some key issues involved in the survey process. One of the earliest surveys—perhaps even the first—is reported in the Bible. After asking Moses to climb Mount Sinai, God commanded:

> Take ye the sum of all the congregation of the children of Israel, after their families, by the house of their fathers, with the number of their names, every male by their polls; from twenty years old and upward. (Num. 1:2–3)

By having Moses take the sum of all the congregation, God was requesting an enumeration of the entire population.

FIGURE 1.1 THE FIRST SURVEY

For many centuries, surveys continued the practice of taking complete population counts, and as societies became increasingly complex, state bureaucracies undertook these data-collecting exercises to learn about the people in the realm. For example, "[t]he Japanese government first compiled population records in [A.D.] 702," surveying both the "composition of the population" and the arable lands using then-current Chinese techniques (Corl and Hayami, 1986: 311–12). Still surviving are documents from the 1500s that record population counts from Arab towns such as Aleppo (Raymond, 1984: 453).

Another example of population counting was a demand made in 1086 by William the Conqueror, the King of England, to compile a list of all landowners in his kingdom. William wanted to establish a national land tax, and the count resulted in the so-called Domesday Book. In the seventeenth and eighteenth centuries in Europe, following years of plagues and wars, concern about the number of people and the food supply led to a steady growth in efforts to gather population counts.

In the late 1800s, the English shipping magnate Charles Booth spent 17 years attempting to find out how many inhabitants of London were poor. Booth relied on "informants," who were generally officials of the state (e.g., police officers, school officials, tax collectors), to provide information. The belief was that more reliable data would come from surveying members of the middle class than from the poor themselves.

The practice of using only a sample of respondents from a larger population is a relatively recent invention. One of the first times it was used was in 1880, when about 25 000 questionnaires were distributed to French workers to study employee exploitation. Among the questions asked were the following:

> (1) Does your employer or his representative resort to trickery in order to defraud you of a part of your earnings? and; (2) If you are paid piece rates, is the quality of the article made a pretext for fraudulent deductions from your wages? (Bottomore and Rubel, 1956: 208)

In this case, the survey researcher was Karl Marx. Unfortunately, no record remains of how many people responded to Marx's survey. Unlike in the earlier examples, Marx did not attempt to contact *all* workers in France. His selection was unlikely to have been representative, but it was a *sampling* of personal opinion he sought, not a complete enumeration. Furthermore, he asked workers to respond for themselves, instead of asking a sample of informants to provide the data as had Booth.

While the wording of Marx's questions seems strikingly biased in light of modern standards of survey research, Marx was probably at least as interested in provoking workers to think about these issues as he was in gathering data. This notion of doing research in an educative manner runs counter to current traditions of scientific survey research, in which data collection is the sole purpose. Nonetheless, many groups (e.g., political parties, sales people) continue to use materials that appear as surveys but whose main purpose is to recruit members, solicit contributions, or make sales.

Another important innovation in the history of survey research can be dated to nineteenth-century England when the so-called Chocolate Sociologists, Seebohm Rowntree and George Cadbury, designed large community surveys to study the nature and extent of poverty. Estimating the extent of poverty is a counting exercise that aims to *describe* the number of poor families. Rowntree and Cadbury's other main focus—to discover the nature of poverty—shifted attention away from simple description to *explanation* and *understanding*. They wanted to know more about the causes of poverty. Their studies introduced new issues of research design and causal analysis into the survey method.

While surveys and poverty have a long historical association, polls and politics are a more recent tradition. Reports of sampling voter opinion date back at least to the 1824 presidential election in the United States (Smith, 1990). In that year, all of the presidential contenders ran under a united Democratic-Republican banner, which made the outcome more unpredictable than it is today under the two-party system. To gauge opinion, newspaper editors relied on straw polls, that is, unofficial sets of votes that were taken at public meetings, militia musters, and social gatherings. As in the case of the survey undertaken by Marx, none of these gatherings truly represented the population of interest (all voters); instead, only a selective segment of the population was surveyed.

In Canadian history, the 1851–52 census was one of the first systematic data-gathering exercises. Interestingly, the theme of poverty and disadvantage was central: "One of the most useful of the Statistical Details connected with the Census is that which points out the number of unfortunate persons who come under the [deaf and dumb] designation" (Census of the Canadas, 1854).

As in other nations, Canadian industrialists played an important role in stimulating survey research. In the late 1890s, inspired by Booth, Sir Herbert Brown Ames (1897), a footwear manufacturer and later an insurance baron, launched a survey of a working-class area of Montreal. Like Booth and the Chocolate Sociologists, Ames was concerned with the social conditions of urban life. Not surprisingly, church groups interested in issues of temperance and moral reform also used surveys. Stewart (1913) reports on two such "general social surveys" conducted in Fort William and Port Arthur (now Thunder Bay) for the Methodist and Presbyterian Churches.

A more recent advance has been the use of survey techniques to gather information about organizations or groups, as opposed to individuals. A survey of colleges and universities might ask about institutional policies on exam regulations for students with disabilities or inquire about student loan policies and procedures. Here is a more complicated example. In the United States, the National Organizations Study draws a sample from U.S. work establishments to ascertain various attributes of workplaces and workers in them (e.g., firm size, pay levels). Obtaining a complete listing of all work establishments is difficult, so a representative list is gathered from respondents to the National Opinion Research Center's General Social Survey (GSS). GSS respondents, who are employees of various establishments, are asked to identify their own place of employment as well as that of their spouse. The responses to this GSS question help generate a list of work establishments (Spaeth and O'Rourke, 1994). This latter example illustrates two points: surveys of organizations, firms, or establishments are now common, and sampling procedures in surveys can be very complex.

As the use of surveys has increased through history, survey techniques have become more refined. The custom of letting people speak for themselves as respondents, rather than relying on informants, was a major advance. The practice of sampling was also a crucial refinement. A complete population enumeration, which still occurs at least every decade in most countries, is now typically referred to as a *census*. In contrast, the term *survey* is usually reserved for studies that employ some sample selection from a population. Another important refinement was the recognition that samples had to represent the full cross-section of the population. Finally, and more recently, computer technology has revolutionized both the collection and analysis of survey data.

SURVEYS IN MODERN CANADA

Today, billions of dollars are spent annually on collecting survey data. Surveys are everywhere. Think about the song currently at the top of the charts. How does anyone decide which song is number one? Knowing which songs people want to hear is important because advertising revenues depend on the number of listeners. To define their market and to determine audience ratings, radio stations typically rely on surveys. In fact, a great many organizations depend on surveys.

Governments, political parties, private corporations, trade unions, community groups, and university researchers are all prime users of survey knowledge. The extent to which survey research is used for basic information gathering is not generally recognized. For example, few Canadians realize that the unemployment rate, which is released through the media every month, is actually determined using survey research.

Every month, Statistics Canada asks a sample of about 52 000 representative Canadian households a series of questions about work. One question that can be paraphrased briefly asks the following for all household members aged 15 years or older: "Have you been working or looking for work in the past four weeks?" The monthly Canadian unemployment rate is then calculated as the number of people who report they are looking for work divided by the total number of people working or looking for work. Although most Canadians know the approximate unemployment rate in the country, few could describe how this rate was determined (most people mistakenly believe that it is a count of people receiving employment insurance benefits).

Another example of how commonplace social surveys have become can be seen in the legal system with the rapidly growing use of survey evidence, especially in the areas of trademark infringement, advertising claims, and jury selection. To cite one case, an established maker of teddy bears took a new manufacturer to court, claiming that the new manufacturer's teddy looked too much like the established maker's teddy, thereby confusing consumers. However, when surveyed, a sample of children had no difficulty distinguishing between the two teddies, and the trademark infringement claim was dismissed. Manufacturers have also been taken to court over allegations of misleading advertising, and survey evidence has been used both to support and to refute these allegations.

Recent and rapid changes in technology have influenced the use and conduct of social surveys. The Web is an obvious example. Survey firms were also quick to adopt the personal computer (PC) for use in designing questions and surveys, as well as for data entry and analysis. The PC has also aided in random dialling of telephone numbers to contact respondents for the survey. Connecting PCs to the Internet led to the development of Web-based surveys. Simultaneously the widespread use of cellular phones has made telephone interviewing more complicated. Technological change has both enhanced survey practices (e.g., computer-assisted telephone interviewing) and complicated survey implementation (e.g., caller ID, confusing surveys with spam on the Internet).

CREATING A GOOD SURVEY IS TOUGHER THAN IT SEEMS

Users of surveys require solid, reliable data. The cliché "garbage in, garbage out" applies to surveys. If a survey contains the wrong questions, good decisions cannot be based on the resulting information.

All of us have experience in trying to get information from others. We ask other people for directions to locations, for their opinions, and for instructions on how to do things (e.g., with class assignments—when they are due, if they should be typed, what the penalty is if they are late, and so forth). We consider

FIGURE 1.2 USING SURVEYS

ourselves experts in asking questions. Consequently, many people believe that constructing a survey is an easy task, since it is often thought to be nothing but a listing of questions. We hope to disabuse you of this notion and to demonstrate how much care needs to be taken to create a survey that will produce significant and meaningful results.

Research can go astray at many junctures. A common mistake early in the process comes in not establishing clearly what needs to be known and why. Following an ill-considered hunch or a vague idea can lead to disaster. More surveys fail for this reason than for any other. Asking the wrong questions usually results from not having thought out beforehand exactly what needs to be known. Thinking first helps to ensure two things: a clear definition of the research question or problem, and an understanding of precisely what information is needed to answer the question. The first five chapters of this book concentrate on the design of good research questions.

Once the problem has been defined, the next step is to formulate well-targeted survey questions. Unless extreme care is taken at this point, it is easy to ask questions that respondents are unable—or may refuse—to answer or ones they perhaps do not even understand. To put it succinctly, good questions elicit good information. The words used and the order in which they appear affect how people respond. Small changes can make big differences. This is the focus of Chapters 6 through 10—asking good questions about relevant issues.

Administration and implementation also influence the quality of data obtained. Getting the necessary information can be facilitated by organizing the survey process effectively. Typically, survey information is used to make conclusions about a large population. This involves issues of sampling (How representative are the respondents? How were they chosen? How many people refused to participate in the survey?), which are covered in Chapters 11 and 12.

Producing a good survey is not only a matter of collecting good information. Considerable attention must also be devoted to the task of organizing the information so that it can be properly analyzed. This is done by entering the answers of survey respondents into computers to take advantage of the computer's capabilities for systematic storage and analysis. Finally, survey researchers must deal with the presentation of results. The final chapter deals with these topics—analyzing and presenting survey data.

In each chapter, we have organized the material so that you can observe the sequential construction of a survey from beginning to end. And since learning by doing is an effective way to master survey skills, we have also devised a series of exercises that will conclude each chapter and provide a clearer understanding of how to develop, design, and carry out an effective survey.

EXERCISE 1.1 IDENTIFYING THE PURPOSE OF A SURVEY

1. Clip an example of a survey from a newspaper or magazine. Identify the purpose of the survey, summarize the key findings, and if possible, describe the sample.
2. Find a scholarly article or chapter segment dealing with censuses or surveys. Write a summary describing the article.

Some of the following sources may be helpful:
Public Opinion Quarterly, International Journal of Public Opinion Research, Journal of Marketing Research, Journal of Consumer Research, American Journal of Political Science, Canadian Journal of Political Science, Gallup Reports, New England Journal of Medicine.

You can also try some of these websites:

- Statistics Canada: http://www.statcan.ca
- The Gallup Poll: http://poll.gallup.com/
- Environics: http://www.environics.ca
- The Roper Center: http://www.ropercenter.uconn.edu
- National Opinion Research Center: http://www.norc.org

FURTHER READING

Babbie, E. (1990). *Survey research methods* (2nd ed.). Belmont, CA: Wadsworth.

Bulmer, M., Bales, K., and Sklar, K.K. (1991). *The social survey in historical perspective, 1880–1940.* New York: Cambridge University Press.

Czaja, R., and Blair, J. (2005). *Designing surveys: A guide to decisions and procedures* (2nd ed.). Thousand Oaks, CA: Pine Forge Press.

Fowler, F.J. (2002). *Survey research methods* (3rd ed.). Newbury Park, CA: Sage.

Jackson, W. (1988). *Research methods: Rules for survey design and analysis.* Scarborough, ON: Prentice Hall.

Marsh, C. (1982). *The survey method.* London: Allen & Unwin.

Rea, L.M., and Parker, R. (2005). *Designing and conducting survey research* (3rd ed.). San Francisco, CA: Jossey-Bass.

Rubenstein, S.M. (1995). *Surveying public opinion.* Belmont, CA: International Thomson Publishing.

Chapter 2

THE PURPOSE OF SURVEYS

Research involves gathering specific information to answer carefully designed questions. Doing this well requires meeting standards that facilitate the systematic and verifiable collecting of information or evidence. The findings from good research allow us to make factual as opposed to conjectural statements about the world—that is, statements based on evidence as opposed to statements based on suppositions or hunches. Survey research is no different. The goal is to provide information (data) by systematically asking questions. Clearly establishing the purpose of those questions is a first step in carrying out good research.

THE PURPOSE OF SURVEYS

As with all research findings, information from surveys is used either to describe, explain, or influence some phenomenon. The information from surveys may be used to:

Purpose	Example
Determine a single value	"How many people would vote for Candidate A, if the election were held today?" In this case, there is only one value—the number of people likely to vote for Candidate A.
Describe a variable—it has more than one value	"If the election were held today, how many people would vote for Candidate A as opposed to Candidate B?" The number of people likely to vote for A versus B.

Describe a relationship (between variables)	"Does either candidate A or B receive a higher proportion of votes from male voters than from female voters?" Members of which gender are more or less likely to vote for which candidate?
Explain a relationship	"Why does one candidate receive a higher proportion of votes from male voters than from female voters?" What makes women (or men) prefer one candidate over another?
Influence something	"How can we increase the level of support for Candidate A among female voters?" What can we do to change things?

Sometimes only one purpose is involved, but frequently each purpose plays a role in the ultimate use of the survey findings. In the latter case, these purposes can be usefully understood as sequential steps that build to a strong research design.

HOW THESE STEPS BUILD ON EACH OTHER

Here is a hypothetical example.

Step 1 A business client may start out saying, "I want to know how many people might use our new product." The client wants to *determine a single value*—the number of people who would buy the new product. That value might be 320 000, the number of people who would buy the client's product.

Step 2 After further discussion the client may say, "It would be useful to compare the number of people using our product as opposed to Zing—the only other product on the market." The client wants to *describe a variable* (type of product purchased). Here the client wants to know more than "how many people" would use the client's product (step 1). The client also wants to know the proportion of people using either product A or B (i.e., market share). In this case the answer would have two values: the proportion of people using the client's product and the proportion of people using Zing. In step 2 the focus switches to the variable "type of product purchased" (a variable because it has two or more values).

Step 3 A few minutes later, the client says, "I would like to know if use [or not] of our new product (and Zing) is related to age [or income or whatever]." The client now wants to *describe a relationship* between two variables: age of the customer and whether or not the customer uses the client's product or Zing. (The concepts of value, variable, and relationship are more fully discussed in Chapter 4.)

Step 4 Thinking further about this, the client then says, "I would like to know why the use of our new product is related to age [or whatever]." Now the client wants to find out why age is related to product choice; the intent is to *explain the relationship*—why does age influence the decision to use a product?

Step 5 Finally, the client wants to know what to do to alter market share: "What can I do to increase demand for our new product? What changes in, for instance, the product, the price, or the advertising will change demand?" The question now becomes: if we did X, what would happen? The client wants to *influence or change something*. For example, "How could I get younger (or older) people to buy our product?"

Each step adds complexity to the information a survey researcher must collect. Step 1 is simple—determining how many people might use this new product. Step 5 involves determining if people would alter their habits in response to changes in price, product design, or advertising.

Below are two different examples illustrating the basic ideas developed in steps 1 through 5. First we provide a statement that you might read in a newspaper or in a research study, then we suggest the kind of question that might have led to the particular statement, and finally we specify the single value, variables, and/or relationships implied in both statements.

Statement 1: Few students at the school agree with the proposed exclusive beverage purchasing policy.

Question: How many students at the school agree with the proposed exclusive beverage purchasing policy?

Single value: The number of teenagers agreeing.

Note in this example that the concern is not about how many students disagreed, or didn't care, or were neutral. The issue was how many agreed. Even though the researcher may have intended to imply a variable (i.e., compared to those who didn't agree), the variable is not explicit. For the purposes of the survey, the number agreeing is a single value (e.g., 42).

Statement 2: More women than men over age 65 live alone.

Question: What are the living arrangements of women and men over age 65?

Variables: Gender (women and men); living arrangements (alone, with spouse, with other relatives, etc.)

Relationship: For seniors, do living arrangements differ by gender?

FIGURE 2.1 STEPS IN DEFINING THE RESEARCH PROBLEM

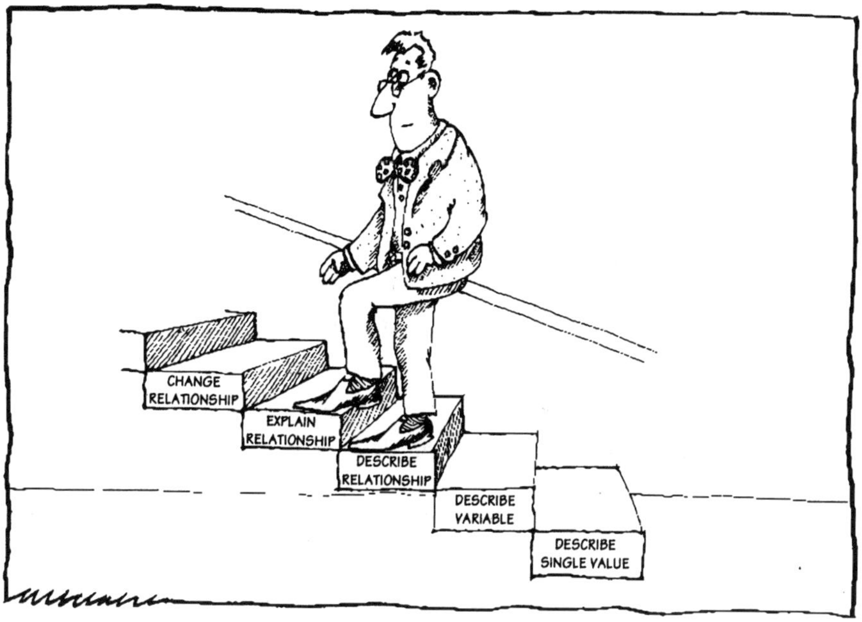

Note here that age is not a variable, but simply a factor restricting the age group to which the stated relationship applies. This is referred to as a *scope condition* (see Chapter 5).

CLARIFYING THE STUDY OBJECTIVES: PLANNING THE COMPLETE PROJECT

Survey researchers can save themselves time and money by recognizing the building steps involved in the research process. At the outset, it is often not clear what exactly a study is meant to accomplish. A major part of the researcher's task is to clarify the survey objectives. This can be accomplished by exploring a series of questions:

Q. 1. What needs to be known?

Q. 2. Why does it need to be known?

Q. 3. What else needs to be known? Why?

Q. 4. What will the results of the survey accomplish? What decisions will be based on the results of the survey?

Q. 5. Is a survey the best method?

Notice that an answer to any one question may require reconsidering answers to other questions. While research typically begins by addressing what needs to be known, the answer to this question is often clarified when we ask why we need to know something or what else needs to be known.

Not all the above questions may need to be asked, but they are tools to help the researcher determine the study's objectives. Remember that more than one study objective may exist. As well, some study objectives may be hidden. Objectives that often go unstated include finding support for a particular decision; finding that a certain product, program, or decision is not supported; completing a successful survey; or having a polished presentation of results.

AN EXAMPLE OF HELPING TO CLARIFY THE OBJECTIVES

Let us sit in on a meeting between a survey researcher and a client, the administrator of a municipal parks and recreation department.

Researcher: What information do you need from the survey? (Q. 1 p.13)

Client: We need to determine the number of residents using our recreation facilities (*describing a single value*).

Researcher: Why do you need this information? (Q. 2 p.13)

Client: We need to determine how effective we are in serving our residents. Put very simply, the greater the percentage of residents using our facilities, the more effectively we are serving our constituents.

Researcher: So we need to count the number of users as well as the number of nonusers (*describing a variable*).

Client: Yes.

Researcher: What else would you like to know? (Q. 3 p.13)

Client: We would like to know some of the demographic characteristics of the users: age, gender, residential area, occupation, education, and so forth.

Researcher: Why is that?

Client: Well, we would like to know what differences exist between users and nonusers of our facilities (*describing a relationship*). We need to know if only certain groups of residents are served by our facilities: for example, the young but not the elderly.

Researcher: Okay, I can understand your concerns. What led to these concerns, and what led to your decision to undertake a survey? (Q. 4 and Q. 5 p.13)

Client: Well, we have been getting criticism from some groups in the community that we are not providing a broad enough range of activities for the residents. Municipal Council seems to be listening to these criticisms. If we find that some major groups are underrepresented in the use of our facilities, then we need to find ways to increase their participation.

Researcher: Let me see if I understand the problem correctly. You think that you may not be providing for certain categories of people in your community. You want to establish if this is true or not, and if true, you need to know what to change in order to increase use by these groups of municipal parks and recreation facilities. (Q. 4 p.13)

Client: Precisely.

The above approach allows the researcher to formulate exactly what the research needs are. It moves through the following steps: (1) description of a single value, (2) description of differences (a variable), (3) what the differences are related to, (4) why the differences are related to these other variables, and finally (5) what can be changed to increase demand. It is important to clarify exactly which step you or the client are aiming to attain. There is no need to be too ambitious, but it is equally important to ensure the survey results will answer the questions of most interest.

SOME SURVEYS NEED ONLY DESCRIPTIVE INFORMATION

Some types of surveys require only descriptive information. For instance, to estimate who is winning the election, exit surveys at polling stations may need only to ask people for whom they voted, while surveys on the "grand opening" of a store may need only to estimate the anticipated size of the crowd in order for the store to make preparations. However, many surveys are carried out with the intention of affecting some behavioural or attitudinal change, whether in purchaser or user demand or in increased support for a program or candidate.

SUMMARY

Clearly defining the purpose(s) of a survey is a necessary first step in the process of carrying out a successful research project. The purpose of a survey can range from describing some phenomenon, to establishing relationships between variables, to providing information about what change(s) can be introduced to produce a desired change in something else. Many people do not clearly establish why they need the survey information. One of the roles of the survey researcher is to help define the purpose(s) for which the survey information is to be used.

EXERCISE 2.1 IDENTIFYING RESEARCH PURPOSES

The following statements are like those often found in newspapers, magazines, and research publications. Your first task is to reformulate each statement as a question. As shown on pages 12 and 13, what question underlies each of the statements? (There are several different ways to pose these questions, so choose a plausible question phrasing.) Second, after posing the question, identify any of the following elements in the question: a single value, a variable or variables, and any relationship. Be careful not to change the meaning of the statement as you specify the applicable elements. Also note any factors explicitly referred to that limit the scope or range of the statement's applicability (as in statements 7 and 9 in particular).

1. The size of the crowd was estimated at 50 000 people.
2. Thirty-two percent of young people smoke.
3. More people drive faster than the speed limit than do not.
4. It is estimated that the local market for formless bagels will be about 73 000.
5. The larger the number of issues in an election, the lower the voter turnout.
6. An individual's status ranking in a group affects the degree to which she or he sets the norms for clothing styles.
7. Within housing projects, the lower the proportion of friendships, the lower the use of common recreation facilities.
8. A participant's socioeconomic status affects whether he or she engages in an individual or team sport.
9. Men attribute successful outcomes of difficult intellectual tasks to their abilities, whereas women attribute the same outcomes to luck.
10. Household recycling effectiveness can be improved by supplying people with more detailed instructions on how to recycle properly.

EXERCISE 2.2 DEFINING A RESEARCH PROBLEM

1. Define a simple research problem. The problem should focus on something specific to the field in which you are specializing (be it political science, recreation, sociology, marketing, planning, health sciences, and so forth), and it should refer to something in which you have had some experience. This

problem is used in later exercises, so select something that interests you, is manageable, and has some connection with your field.

Examples:

a. Why do more women enroll in fine arts courses than do men?

b. Does the reading of beauty magazines influence women's self-confidence?

c. Are families with children more likely to recycle and compost?

2. Clarify what is meant by the terms you use (e.g., fine arts courses, self-confidence). State your terms as variables (e.g., for the first example, "gender" and "course selection" are the variables).
3. Think about two or three additional factors that may be connected to the variables you have identified (e.g., interest in fine arts, artistic ability).

FURTHER READING

Becker, H.S. (1998). *Tricks of the trade: How to think about your research while you're doing it*. Chicago: University of Chicago Press.

Berk, R.A., and Rossi, P.H. (1998). *Thinking about program evaluation* (2nd ed.). Newbury Park, CA: Sage.

Payne, G., and Payne, J. (2004). *Key concepts in social research*. London: Sage.

Walliman, N. (2005). *Your research project: A step-by-step guide for the first-time researcher* (2nd ed.). London: Sage.

Chapter 3

ETHICAL ISSUES

Moral and ethical dilemmas are common to all professional activities. In occupations such as engineering, medicine, and accounting, strong associations have established ethical standards to help guide practitioners. In contrast, surveys are carried out by researchers from many professional backgrounds (e.g., university-based social scientists, market researchers, health-care methodologists), each with separate ethical codes. Consequently, no single set of ethical principles serves to guide the survey researcher (see Palys, 2003, for excerpts from various ethical codes).

Prior to carrying out research on human subjects, government-sponsored research teams and university-based researchers submit their research proposals to ethical review panels. If the planned research does not meet the standards, funding is withheld or the university does not permit the research to be carried out. In the private sector, these reviews are not present.

Most survey research is undertaken by someone (a researcher) for someone (a sponsor or client). Researchers often work for market research or public opinion polling firms, although they may also be employed in the civil service or in universities and colleges. In these and other research situations, ethical issues involve the relationship of the survey researcher with clients or sponsors, employers, survey respondents, coworkers, research consumers and/or decision-makers, and the general public. Often, obligations to one or more of these groups conflict. For example, research sponsors may need systematic information to help in planning better drug and alcohol programs, but asking respondents questions about their use of drugs and alcohol can be traumatic to interviewer and respondent alike.

Some moral dilemmas may be addressed by the ethical codes of the different professional associations to which a survey researcher might belong. However, even with professional guidelines in place, ethical decisions ultimately rest with the individual researcher. No set of rules or procedures can cover all of the possible ethical dilemmas a researcher might face, and so personal interpretation and decisions are essential.

We concentrate our attention on two aspects of survey research in which ethics are involved. First, we focus on relations between the researcher and the survey sponsor or client. Second, in the latter part of the chapter, we attend to relations between the researcher and the survey respondent.

SURVEY RESEARCH AND CLIENTS/SPONSORS

A basic ethical dilemma for survey researchers is in deciding how far it is possible to cater to the interests or wishes of a client or sponsor. A substantial amount of survey research is undertaken for third parties, be they private firms hiring market research companies or labour unions hiring polling firms. For example, as we will see in later chapters, it is possible to influence results by the choice of wording in the questions asked on a survey. While in theory it is easy to say that researchers must be as objective and neutral as possible, it is naive to think that no outside influences shape the thinking of survey researchers, especially when they face losing a few contracts worth thousands of dollars.

A basic ethical dilemma for survey researchers is deciding whether to engage in an artful con or a truthful art. Should the criterion used to judge research information be the *appearance* of solid, believable, and acceptable results or should the criterion be acceptable methodological *practice*? Makeup artists and advertisers are experts at exploiting appearances and sometimes gloss over fundamental weaknesses or problems in the hope that consumers will make judgments based on appearance and not practice. In research, basic attention needs to be paid to methodological practice, and attempts to gloss over problems are dishonest and unethical.

The ethical decision-making process is often critical because most research consumers lack the ability to distinguish between a good and bad survey. They cannot always judge whether the questions measure what they are intended to measure, or whether the sample is representative.

Audiences rely on cues to determine the believability of results, including the presentation of both the survey researcher as an expert and survey research as a scientific practice. Other cues that influence judgments of research credibility include the precision of the reported findings (e.g., 23.726 percent of respondents approve of XYZ) and the visual appearance of the presentation. Clients unwittingly contribute to the ethical dilemma of the researcher by frequently basing decisions on believability factors rather than on criteria of sound research practice.

The following are three typical situations in which believability factors are given priority in the conduct of the survey:

Situation 1: Look, we can't afford to pay too much to do the research. When we go to advertisers, all they care about is whether the survey shows that our readers are in the occupational categories that buy their products.

In the above scenario, the sponsor is more concerned with whether the audience (advertisers) will believe the results than with whether the research is sound.

Situation 2: It may not be the best way or even a good way to get the information, but our mandate clearly stated that we were to confer with community members. People expect a survey.

Here again the actual research results are unimportant. The main concern is that a survey has been done. Often a key criterion is an economic one: "It had better not cost much!"

Situation 3: Don't you think a survey showing consumer satisfaction with our product would be an important part of our presentation package for an extension of our operating line of credit?

In the third example, the client wants a particular result to occur. The interest is not in what actually exists. Entrepreneurs going to the bank for an extension of their line of credit want survey results showing market demand for their new product. A developer may want results showing that the majority of residents approve of a building development.

In the above three cases, survey researchers are faced with a conflict involving the ethics and standards of their profession, their personal ethics, and the demands of the research sponsor. Some survey researchers rationalize the conflict by saying that this is the way the world is and everybody plays the same game. They may come to see making ethical compromises as part and parcel of making a living.

Completing a solid piece of research can be expensive. It may involve considerable work prior to the actual survey, including exploratory studies and the pre-testing of questions. When putting together a proposal for a client who has asked a number of firms for quotations, the surveyor knows that if she or he develops a proposal for good research as opposed to lower quality research, the higher costs of the good research may decrease the likelihood of being selected to do the work. Researchers unwilling to compromise professional standards are thus placed in a tough situation. Does the researcher maintain the standards and possibly lose the work?

FIGURE 3.1 THE ARTFUL CON

A solution to this problem used by reputable consulting and survey research firms is to establish guidelines. Is the client willing to have the limitations of the study and data clearly stated in the report? Is he or she willing to accept the limitations of the study when making decisions? Will the client mention data limitations when making presentations to other parties? Such strategies can help resolve ethical difficulties without compromising the researcher's integrity.

Survey researchers can also unwittingly compromise the quality of the research by ignoring their own professional standards. Researchers can unintentionally take on the values of the client and lose their professional impartiality. They, like many other consultants, become advocates rather than independent experts. The client's biases may creep into the process when the surveyor is developing the research problem, formulating the questions to be used, and deciding which findings to interpret.

USING OTHERS TO MAKE A LIVING

A second major area of ethical concern centres on relations between a survey researcher and the survey respondents.

The Benefits to the Researcher

Researchers make their living using data collected from survey participants. The more willing respondents are to participate, the better it is for the researcher.

FIGURE 3.2 ASSUMING THE VALUES OF THE CLIENT

While researchers cannot coerce people into participating, they will actively encourage them to do so.

Encouragement takes a variety of forms, including offering monetary incentives, employing persuasive interviewers, and emphasizing people's contributions to a worthwhile project. Research firms sometimes use deception to hide both the purpose of the survey and the sponsor of the survey, particularly if they feel that disclosing this information will reduce their ability to attract respondents.

The Costs to the Participant

The respondent commits her or his time to reply to questions that may touch on sensitive issues or arouse negative emotions—with little apparent reward for doing so. Respondents also take the risk of contributing to a study whose results may not be in their best interests. Whether conducted by mail, on the doorstep, by telephone, or using the Internet, survey research necessarily intrudes on people's privacy, even if it asks them to participate voluntarily. Potential costs to the respondent must be weighed against possible benefits to researchers.

The Invasion of Privacy

The increasing use of surveys by governments, businesses, and nonprofit organizations has resulted in more intrusions into people's lives. Most participants complain about telephone surveys. Unlike mail or Web-based surveys that can be thrown away, ignored, or viewed at one's convenience, telephone surveys directly interrupt the activities of the people contacted. If the potential participant answers the phone, she or he will likely spend at least a few minutes listening to the opening statement and deciding whether or not to participate. The most advantageous time for survey firms to reach respondents is in the early evening hours, but many people resent having their dinner or their leisure time interrupted.

Internet surveys have also been criticized as an invasion of privacy and labelled as unwanted spam or obnoxious e-mail messages. As with the telephone survey, participants in Internet surveys cannot easily check the legitimacy of the surveyor. In addition, the possibility of matching the respondents' identity with the survey data might put participants at risk.

Weighing Researcher Benefits and Participant Costs

Since it is easier for researchers to see the benefits to themselves rather than the costs to participants, the codes of ethics of professional associations and ethical review committees of many organizations provide safeguards to protect respondents. Researchers must at all times try to minimize the costs to the respondent and treat people with sensitivity. In practice, a cost-benefit comparison is very difficult because typically the actual costs and benefits are unknown. Researchers are thus in a very difficult position when it comes to weighing their interests against the interests of others.

Informed Consent

People must be made aware that their participation is voluntary, and they must have enough knowledge about a study to give researchers informed consent prior to participation. This means that at the very least they must be told about the overall purpose of the research and the general content of the questions. In addition, respondents should have an opportunity to ask questions before an interview begins. In the case of a postal survey, a telephone number ought to be provided to allow respondents to ask questions or to verify the legitimacy of the survey. As we discuss in later chapters (particularly Chapter 10), respondents must be provided with information that allows them to make a reasoned decision about their participation.

A Question of Ethics

Researchers need to ask themselves several questions about ethical issues involved in a study. As well, researchers often have to present their proposals to ethical review panels. Some of the questions to be addressed are as follows:

1. Who will conduct the study?
2. What qualifications do they have?

3. What do you know about the risks of the study?
4. What do you know about the possible benefits of the study?
5. Are the participants able to give informed consent, or are there any physical, mental, or linguistic barriers?
6. Are the subjects likely to sustain long-term discomfort because of the study?
7. Deception compromises informed consent. Does your study include deception? If so, how do the benefits of the study outweigh the costs to the participants?
8. Will there be any long-term or permanent damage to the participant as a result of the deception in this study?
9. What kind of (and how much) compensation, if any, is being offered to the participants?
10. How do you plan to maintain confidentiality and anonymity? And will any information that identifies individuals be available to anyone?
11. Who will have access to the results of the study?
12. How do you plan to destroy the data collected?

Figure 3.3 contains a sample consent form that illustrates the details respondents need in order to make an informed decision about their participation.

FIGURE 3.3 SAMPLE CONSENT FORM

Department of Communications
University of Elsewhere

Informed Consent Form

"The Effects of Depictions of Violence Against Women in Mainstream Films on Young Women's Sense of Safety"

Principal Investigator:
Dr. Jane Brown, Department of Communications (222) 555-1234

Co-Principal Investigator:
Professor Chun Lee, Department of Psychology (222) 555-4321

Purpose: To investigate the effects of imagery of violence against women in mainstream films on young women's perception of physical safety.

(*Continued*)

Study Procedures: As a study participant you will be required to view several recent, popular films that include violent imagery directed at women. You will view these films over the course of a few weeks. You will complete a number of questionnaires that include questions about your sense of physical safety, personal experiences with physical violence, and your opinion about violence.

You have been randomly sampled from a list of all undergraduate female students currently attending the university.

Confidentiality/Anonymity: Information gathered from this study will not be traceable to a specific respondent. All identifying markers on all documents will be removed, and participants will not be identifiable in any reports. All responses will be kept anonymous and confidential.

Remuneration/Compensation: All participants will receive an honorarium of ($X) to defray the costs of any inconvenience participating in the study may cause, such as loss of studying time, wages, or the cost of additional transportation.

Contact: If you have any questions or concerns regarding the study, please contact me, Dr. Jane Brown, at (222) 555-1234, or one of my associates at (222) 555-6789.

If you have any questions or concerns about your rights as a research participant, please contact the Director of Research Policy and Procedures at the University of Elsewhere, Dr. Anna Rodrigues, at (222) 555-9876.

Consent:

I understand that participation in this study is completely voluntary. I have the option to withdraw from this study at any time, without risking any adverse effects to my status as a student or any aspect of my student life. I will not be penalized in any way for refusing to participate or withdrawing from this study.

I have read and understood the Informed Consent Form with the Purposes and Procedures of this study. I have received a copy of this consent form, and will keep it for my records. I understand that my anonymity and confidentiality will be protected to the best of the ability of the researchers.

I, the undersigned, consent to participate in this study.

____

Signature of Subject Date

____

Signature of Witness Date

____

Signature of Researcher Date

THREATENING QUESTIONS

Sometimes surveys will include questions that respondents may find psychologically threatening. If there is a potential for harm to respondents, the best policy is often not to carry out the survey. However, occasions arise where a survey is the best method for collecting information, despite the threat to respondents.

For example, in response to the fact that police reports of crimes tend to underrepresent actual levels of criminal victimization, survey research methods have been used to calculate more accurate estimates of criminal victimization. The ethical dilemma for the researcher lies in weighing the benefits of more reliable crime statistics (and potentially a better criminal justice system) against the risks of rekindling personal anguish for respondents.

Other examples of cases where this type of dilemma may arise are numerous, especially if they touch on potentially illegal or immoral behaviour such as toxic waste-disposal methods, child discipline, drug and alcohol use, or white-collar crime.

Handling Threatening Questions

In situations of possible threat to respondents, survey researchers must use trained interviewers who have the skills to help respondents cope with the aftereffects of questioning. For example, researchers who are asking about such sensitive topics as child abuse, substance-abuse problems, or care for elderly parents should provide respondents with information about the topic after the interview. If there seems to be a potential for serious aftereffects, follow-up counselling with therapists should be supplied as part of the study.

For extremely sensitive topics, when only the most skilled and highly trained face-to-face interviewers ought to be used, it is always essential to have a debriefing session immediately following the formal survey interview. As well, it is important to follow up a few days later to ensure there have been no delayed aftereffects. Survey researchers who delve into sensitive or threatening topics must accept responsibility for the consequences of asking people related questions. In thinking about the wisdom of probing certain areas in a survey, it is useful to consider the topics in Figure 3.4.

When a survey contains sensitive or threatening questions, researchers may promise respondents either anonymity or confidentiality. *Anonymity* guarantees that once the survey is completed, no link can be made between the information collected and individual survey respondents. In contrast, *confidentiality* involves keeping secret any link between the information and the respondent by excluding the name, address, or any other direct identifying information from the survey questionnaire. If identification codes are placed on the questionnaires, the key to these codes is kept in a locked cabinet, and procedures

FIGURE 3.4 THINKING ABOUT THREATENING QUESTIONS

Is this a potentially threatening question?

- Yes
- No—include question

Does this question need to be asked?

- Yes
- No—exclude question

Be certain to consider each of the following three issues:

i. What is the best survey format in which to ask the question?

- ❑ Self-administered questionnaire form
- ❑ Face-to-face interview
- ❑ Telephone interview

ii. What kinds of ethical issues should you consider?

- ❑ Assuring confidentiality (e.g., in written reports)
- ❑ Removing all identifying information to ensure anonymity
- ❑ Providing information on the survey sponsor(s)
- ❑ Ensuring proper interviewer training
- ❑ Guaranteeing informed consent

iii. What kinds of resources should you provide for the respondent after the interview?

- ❑ Counselling
- ❑ Hotlines
- ❑ Literature on the topic

are established to control access to the information (such procedures are necessary when respondents are to be re-interviewed at a later date). However, survey researchers have no legal protection if they are asked to divulge records, and while it is rare for courts to make such requests, promises of confidentiality to respondents may be difficult to keep.

HOW ARE THE DATA TO BE USED?

Calculations of benefits and costs are also relevant to the question of how the data are to be used. Are the data going to be used in ways that run counter to the respondents' interests? Will market research data manipulate the customer into purchasing certain products? Will public opinion polls on a major residential development be used to convince city council that the concerns raised during public hearings are not supported by the majority of residents and therefore should be dismissed?

The publication of research results can also threaten individual privacy. If the number of people surveyed is small, or if some kinds of respondents are relatively uncommon, care must be taken that identities cannot be inferred from published reports. For instance, if pulp mills in a province are surveyed about their treatment of dioxins (with the researcher guaranteeing each mill's anonymity), and results are reported by region and size of work force, someone might be able to infer exactly which mill in which region is producing exactly how much dioxin. Or, if you are reporting results from a small neighbourhood or community, and you show family incomes by length of residency and by dwelling type, you may inadvertently identify specific households.

It is also the researcher's job to ensure that the information collected is destroyed properly. While computer data files are often archived for later use, survey questionnaires and code sheets should normally be shredded at some designated point, or at least kept stored in a safe and secure location.

Finally, respondents ought to be informed about how they can access any reports or summaries of the study. Often, it is sufficient to alert them to the intended channels for dissemination.

RESPONSIBILITIES TO SURVEY STAFF

Researchers also have ethical obligations to the staff they employ on survey projects. Two central obligations are training and safety. Survey staff must be properly trained so they can conduct their jobs properly and responsibly. They must know the general purpose of the research and be able to answer any questions that the respondents may ask of them. The information provided in a survey must be treated as private, and interviewers must be cautioned against disclosing any of the information.

The safety of interviewers must also be considered. In face-to-face interviews (and, to a much lesser extent, in telephone interviews), the safety of interviewers may be at risk. The researcher is responsible for developing procedures to ensure the physical safety of interviewers. These procedures can include having the interviewer inform the supervisor of her or his interview timetable, establishing periodic call-ins, sending out interviewers in two-person teams in higher risk areas, and training interviewers to recognize danger signals and to cope with verbal abuse from respondents.

RESPONSIBILITIES TO READERS IN A FINAL REPORT

Certain essential information about the research project should be made public through disclosure in the final report:

1. *Sponsorship.* Who paid for the study?
2. *Timing.* When did data collection begin and end?
3. *Method.* How were the data collected?
4. *Population.* What group do the data represent?
5. *Sample frame.* How was a representative sample acquired?
6. *Sample size.* How many respondents participated?
7. *Response rate.* What percentage of the respondents contacted agreed to participate?
8. *Questions.* What was the exact wording of the questions?
9. *Measurement quality.* How accurate are the data?

Failure to disclose the above information prevents public criticism and compromises the ethical obligation researchers have to their professional peers, to the public, and to research sponsors.

SUMMARY

Conducting a survey involves the researcher in a number of ethical issues that need to be addressed in the planning stage (hence the placement of this chapter immediately after the statement of the research problem). The ethical issues pertain to the relationships of the survey researcher both to the client and to the audience to whom the survey results are addressed. Researchers must remain vigilant about ethical issues while making a living or furthering their careers. In addition, the researcher has responsibilities to the respondents and survey staff that affect how data are collected and how they are used.

EXERCISE 3.1 SENSITIVE TOPICS AND THE SURVEY

Find an example of research on a sensitive topic in which survey information has been used. Examples might include spousal abuse, drug and alcohol use, abortion, criminal victimization, or marital relations. Write a brief summary of the research objectives and comment on the ethical issues—were these adequately described in the writing, and do the respondents seem to have been given fair ethical treatment? What else might have been done, or done differently, to guarantee informed consent and respondent protection?

FURTHER READING

Barnes, J. (1979). *Who should know what? Social science, privacy and ethics*. New York: Penguin.

Homan, R. (1991). *The ethics of social research.* New York: Longman.

Presser, S. (1994). Informed consent and confidentiality in survey research. *Public Opinion Quarterly, 58*(3), 446–59.

On the Internet

American Psychological Association: http://www.apa.org/ethics/

Canadian Sociology Association: http://www.csaa.ca/structure/Code.htm

Canadian Institutes of Health Research: http://www.cihr.ca

Government of Canada, Interagency Advisory Panel on Research Ethics: http://www.pre.ethics.gc.ca/english/policystatement/introduction.cfm

Chapter 4

DEVELOPING THE RESEARCH QUESTION

Successful surveys build on two kinds of questions—the research question and the survey questions. The *research question* is the puzzle, problem, or issue that the research addresses (what do we want to know?). The *survey questions* are the questions asked of respondents; these questions depend on the research question. Defining the research question is both the starting point of the research process and the backbone of the survey, because it directs the researcher's attention to exactly what survey questions need asking. Formulating intelligent survey questions requires spending considerable time and effort developing the research question (e.g., who will vote for candidate X, what families purchase Y, or why do more women than men like music of type Z?). Figure 4.1 illustrates the link between a research question and the survey questions.

FIGURE 4.1 THE RESEARCH QUESTION GUIDES SURVEY QUESTIONS

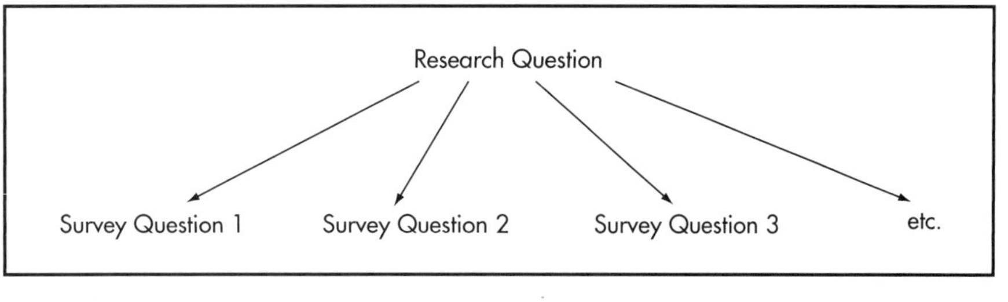

THINKING BEFORE ACTING

Developing the research question begins with clearly stating the issue or problem the researcher or client wants answered. Exactly what do we want to know? Deciding this is far harder than most people realize. As we showed in Chapter 2, the art of asking clear, research-focused questions is difficult.

We can begin the process of specifying the research question by focusing on the key variable that needs to be described, explained, or influenced—the dependent variable.

Two additional elements of the research question are then required. First, *what* variables might affect the dependent variable? Second, *why* might these influences occur? In a nutshell, we must determine what needs to be explained (the dependent variable), what things might do the explaining (the independent variables), and why we think the latter will affect the former.

Alternatively, we can begin the process by focusing on an independent variable, particularly when we are evaluating an action or a program. The research question becomes: What impact is this action or program going to have? For example, a university might introduce a computer program that checks student essays against publicly accessible Web pages. Then the questions would be: Will this reduce plagiarism? What other impacts might it have?

Asking a clear research question is the most important step in doing good research. Research questions need to be constantly refined and ideas continually clarified until the exact nature of the problem is nailed down. Initially, we need to "break the problem open." The process begins with the researcher's own experience and ideas, then moves to the use of other research and writing found in the relevant literature. The goal is to construct a parsimonious model summarizing the research question under examination.

The process typically begins with an initial statement of the research question in which the dependent variable is identified and defined. Next we address key variables that influence the dependent variable—that is, the independent variables. Here we pay attention to causal connections by providing a rationale for how the links we are postulating might actually work (e.g., how does gender affect employment earnings?). Drawing a schematic model of the relationships between variables helps, especially in thinking through how the independent variables might themselves be related to each other. By moving through the steps (outlined in Figure 4.2) in developing a research question, researchers clarify the fundamental issues the survey must address.

STEP 1: STATING THE RESEARCH QUESTION

The first step in any survey is to identify the problem to be investigated—what it is we want to know. Poorly formulated research problems result in bad

FIGURE 4.2 STEPS IN DEVELOPING A RESEARCH QUESTION

1. Stating the research question.
2. Identifying the variables in the research question.
3. Defining variables.
4. Specifying the independent and dependent variables.
5. Thinking causally.
6. Embedding a rationale in causal thinking.
7. Relating the research question to the literature.
8. Introducing other variables (antecedent/intervening).
9. Revisiting causal connections.
10. Drawing a causal model.

research because vagueness undermines the survey process. Without a focus, there are no guidelines for deciding what survey questions to include or exclude. Clarifying the research problem is the most critical stage for a successful survey.

Defining the research question starts with a statement of what we are trying to describe, explain, or influence. In a voting poll, the question may start with wanting to determine the number of votes a candidate is likely to receive. In a marketing survey, the question may start with wanting to describe the product consumers buy in a particular type of market. In a health survey, the question may start with wanting to describe people's attitudes about exercise or nutrition.

Some researchers use the language of hypothesis or proposition to talk about what we are calling the research question. These differences are largely semantic at this stage.

As our example in Chapter 2 about residents' use of recreation facilities showed, you usually begin with a fuzzy statement of the question. This fuzziness is reduced by asking a series of questions of yourself, your colleagues, and your clients. You brainstorm. You consult experts. You talk with associates. You attend to any misunderstandings about the research purpose by streamlining your ideas and making them clearer and sharper. Just as a thesis statement is important in writing a good essay, a crisp problem statement is essential for a successful survey.

Here is how a research question can be refined in a series of successively more focused statements.

1. I would like community feedback on our hospital development proposal.
2. I would like to know how many community members support our hospital proposal.
3. I would like to know what percentage of community members support our proposal.

4. I would like to know who supports, and who opposes, our proposal.
5. I would like to know the characteristics (e.g., age, income, marital status) of the people who support our proposal compared to those who are opposed.

Focusing a research statement becomes a careful balancing act because you want to maintain a research question of manageable size, but you do not want it to be trivial. Like writing a good essay, preparing the initial research question involves continual polishing as you clarify exactly what the research will—and will not—accomplish. While you are developing a research question, the original intent can sometimes be distorted, so be sure that the question you end up posing is the one to which you really want an answer. An initial statement of the research question can be developed in many different directions, thus changing what is to be studied.

	Example 1	**Example 2**
Initial Statement:	I am interested in voting patterns in electing federal politicians.	I am interested in voting patterns in electing federal politicians.
Restatement 1:	I am interested in how people vote in federal elections.	I am interested in how people choose who to vote for.
Restatement 2:	I am interested in whether there is a pattern to how people vote in federal elections.	I am interested in those factors that influence which candidate people choose.
Restatement 3:	Do people vote consistently for the political party rather than the candidate?	Do factors like a candidate's appearance, dress, etc., affect the likelihood of people voting for them?

Although both examples started with the same initial statement, a minor change in wording in the first restatement (how people vote versus how people choose) very quickly resulted in major differences between the final research questions. It is easy to continually change the statement of the problem by changing words. How does the researcher guard against unwittingly changing the meaning of the research question? One way to maintain a consistent definition of the research question is to clearly establish the concepts as variables.

STEP 2: IDENTIFYING THE VARIABLES IN THE RESEARCH QUESTION

After determining our research question, we define it in "variable" format. Variables are measurable concepts that have a range of possible values. For example, length of community residence is a concept we can measure by asking about the number of years someone has lived in a particular community.

Schools have different numbers of students; students have different ages. The number of students and student age are different variables related to schools and students. Students' different ages are the values of a variable, age. It is useful to distinguish between *concepts* as ideas or abstract constructs, and *variables* as measurable referents of concepts (see Chapter 6).

In the example above about support for the hospital development proposal, we might be interested in knowing whether long-term residents of the area are among the main supporters. There are two variables here: level of support for the proposal and length of residence in the area. People in the local community will vary both in their level of support for the proposal and in how long they have lived in the community. "Level of proposal support" is one variable, and "length of community residence" is a second variable.

STEP 3: DEFINING VARIABLES

Producing an exact definition of variables is not always so easy. For example, leisure time or patient satisfaction are concepts that can mean different things to different people. Before creating questions to measure variables, we must clarify what variables mean.

A starting point in defining a variable is to think of examples that reflect its different values. For example: married, separated, divorced, and never married are some values of the variable "marital status." The values of a variable will need to capture the full range of the variable and be mutually exclusive categories.

In researching public transit ridership, a researcher may have in mind some different types of ridership (i.e., some different values): people who use public transit as their sole means of transportation; people who commute to work using public transit, but who use a car for other travel purposes; people who use public transit only when their cars are being serviced or repaired; and people who never use public transit. But what exactly are these examples capturing? What variable or variables underlie these categories? One variable might be the frequency with which people use public transit. A second variable might relate to the reasons people have for using public transit (e.g., commuting to work). A third variable might relate to car ownership.

Concepts like "transit use" or "leisure time" are simplifying phrases used to summarize abstract ideas. All research questions contain concepts—that is, shorthand expressions of some phenomenon in which a researcher is interested. Clarifying research questions requires providing details on exactly what these concepts or shorthand expressions actually mean (e.g., what exactly is "leisure time"?). Concepts must be related to empirical examples. If the concept "leisure time" is part of the research question, what does the researcher mean when he or she uses that expression? Activities such as reading, working out at the gym, doing the cooking, or shopping for clothes might qualify as leisure pursuits for some people, but not for others.

The following six guidelines can be used by researchers to clarify their concepts:

1. *Search for alternative meanings of concepts.* Examine how others have defined terms. Collect as many different definitions as possible. How have others defined "leisure"?
2. *Include the full range of meaning in concepts.* Determine the core features of a definition. Be sure that these features incorporate the entire realm of a concept. For example, "basic mathematical ability" must include division along with the other mathematical skills since it is a key part of mathematical ability.
3. *Be sensitive to various dimensions of concepts.* Many concepts have several aspects or subcategories. Identify various dimensions explicitly. For example, "job satisfaction" has many components: amount of pay, degree of boredom, level of exertion, relations with coworkers, extent of supervision, and so forth.
4. *Be wary of combining two or more concepts.* Distinguish between separate dimensions and separate concepts. Very general concepts like job satisfaction may include more specific concepts such as economic rewards (pay, benefits, perks), job stimulation (challenge, responsibility, innovation), and so forth. Establish the right level of abstraction.
5. *Be sensitive to the extremes of a concept.* Do you want to know about the degree of something, or are you interested in polar opposites? Support for conservative values ranges from low to high, but conservatism can also be contrasted with socialism. Consider opposites. Concern for job *dis*satisfaction helps clarify job satisfaction.
6. *Avoid circular logic in defining multiple concepts.* Beware of creating artificial causal connections between concepts by definition. For example, "survival of the fittest" can be true by definition if the fittest are defined as those who live the longest (i.e., those who survive). Definitional truths are trivial and uninformative. They are true by definition and do not add anything to our knowledge.

These guidelines help in defining precisely what a concept means (and does not mean). Beyond defining terms, we must also measure them—that is, we must establish empirical indicators for concepts. Although measurement is not discussed until Chapter 6, it is important to remember that eventually survey questions have to be asked to measure all of the variables included in the final theoretical model.

Levels of Abstraction and the Research Question

Concepts used in research questions range from the general to the specific. For example, the concept of "political choice" is very general, while the concept of "candidate voted for" is much more specific. The concept "social ties" is more general than the concept "friends," but the concept "employment earnings" is more specific than the concept "income." This is called the *level of abstraction*, with general concepts being more abstract than specific concepts.

Many research questions are difficult to develop because they are stated either too specifically or too generally. If you have trouble developing a research question, try to either broaden it by making concepts more general or restrict it by focusing on more specific concepts. A research question on property crime among young people can be made either more general—criminal activity among people over 10 but under 30—or more specific—break-and-enter incidents among juveniles. Achieving a good balance in the level of abstraction is important. You want to be neither too specific nor too general, as demonstrated in Figure 4.3.

FIGURE 4.3 LEVELS OF ABSTRACTION AND A GOOD RESEARCH QUESTION

Einstein is reputed to have said that asking the right research question takes you 90 percent of the way to doing good research. Balancing simplicity and complexity is crucial in asking a good research question. Be careful not to make your research question too broad and unmanageable. But also be wary of the opposite danger—making your research question too narrow and simplistic. Moving your research question up and down through levels of abstraction will help you focus.

Too narrow

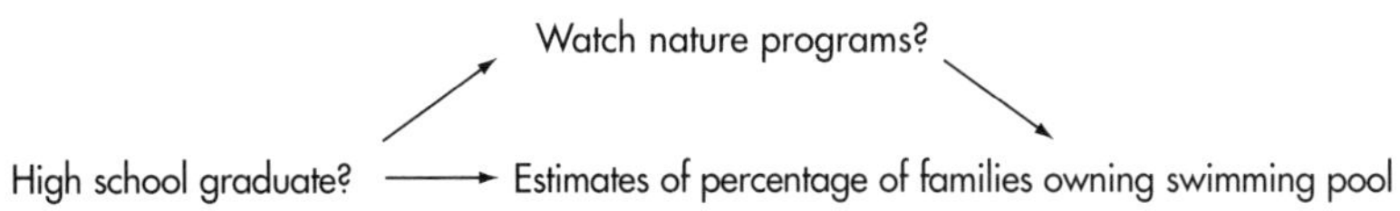

Too broad

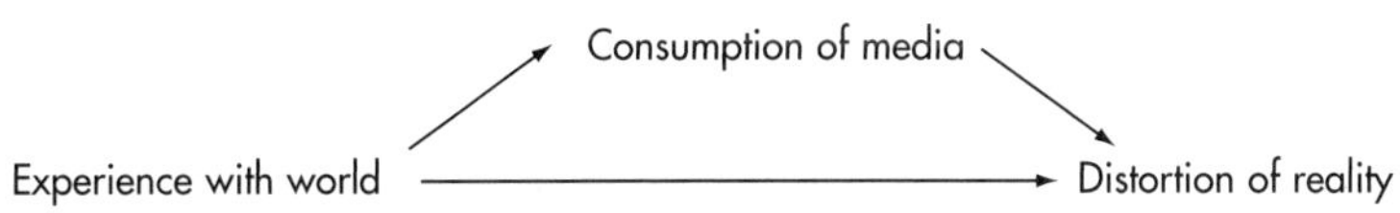

STEP 4: SPECIFYING THE INDEPENDENT AND DEPENDENT VARIABLES

Most research questions have only one variable that we are trying to describe, explain, or influence. If our quest is to describe something, we have a very simple research question. How many people will vote for each candidate? Typically, however, we want to know something more complex. Are women rather than men more likely to vote for one candidate instead of another? In this research question we are asking whether gender affects choice of candidate. Stated another way, we are asking whether choice of candidate depends on the gender of the voter. We are not assuming that gender completely determines whom you vote for (i.e., that women vote only for candidate A and men for candidate B), although this may happen. Instead, we are asking whether the likelihood of a voter choosing candidate A or B is related to the gender of the voter (are women more likely to vote for candidate A?). In this example, candidate choice is the dependent variable, gender is the independent variable.

In deciding whether one variable depends on another, we are thinking causally. We are saying that one variable causes another variable. You have lots of experience thinking causally, but you need to concentrate on how to apply that thinking to research questions. It is common to speak of the cause of an accident or the cause of an illness. In these two examples, the accident and the illness are equivalent to dependent variables. They might have depended on speeding (an independent variable affecting the likelihood of being in an accident) or of not eating healthy enough food (where the type of food eaten is an independent variable influencing health). The independent variable causes the dependent variable. Remember, however, that we are not claiming that one variable completely determines another, merely that one variable influences or has an effect on another. Many people eat unhealthy food and do not become ill, and conversely some people eat only healthy food and nevertheless become sick. What we eat does not completely determine the status of our health.

STEP 5: THINKING CAUSALLY

For something to causally influence something else, four conditions must be met. First, there must be some relationship between both variables—in other words, they must co-vary. This means that change in one variable must be related to change in another. For example, as we age we tend to become more and more sedentary, engaging less and less in physically exhausting sports. Change in age varies in a consistent way with levels of exercise. However, age does not co-vary with listening to music. Young people listen to music as much as older people (although musical tastes change with age). Age does not influence musical consumption; since the two are unrelated, they do not co-vary.

(As an important aside, notice that there are individual exceptions to all of these claims. Some older people do engage in physically demanding activities and some young people do not listen to music. However, as survey researchers our interest is in patterns of behaviour, not in exceptions or rare experiences.)

Second, the variable that does the influencing must occur or change before the variable that is being influenced. Temporal ordering is important, and the independent variable must occur prior to the dependent variable.

As a third condition of causality, we need to be on the lookout for spurious causal linkages. A favourite example is the number of fire trucks attending a fire and the amount of fire damage. Fires that result in a lot of damage do attract a lot of fire engines. These two variables (number of trucks and amount of damage) do co-vary. But this relationship is not causal because we know there is a third variable, the size of the fire, that causes both the number of trucks responding and the amount of damage from a fire. To say the original relationship is spurious is not to say it is false; rather, it is to say that the original relationship is causally misleading. To think causally requires you to think about other factors that might be influencing the relationship between any independent and dependent variable.

The fourth, and final, condition for causality is the clear statement of an explanatory rationale for the causal link. We develop this idea as step 6.

STEP 6: EMBEDDING A RATIONALE IN CAUSAL THINKING

To be effective in thinking causally, you need to develop a rationale explaining how it is that one factor might be related to a second. The debate over smoking and cancer nicely illustrates this issue. For a long time, cigarette manufacturers were able to win costly lawsuits because, although smokers were more likely to develop lung cancer than nonsmokers, the process by which smoking caused cancer was not understood. There was a correlation between smoking and cancer (they co-varied), but that correlation did not prove causation. To be able to say that smoking caused cancer, we had to know how the influence worked. Medical researchers have now made this link by showing the pathway by which the inhalation of smoke causes cells in the lungs to mutate and become cancerous. Similarly, we can be comfortable saying that aging causes a reduction in physical activity because we know that as we age our flexibility lessens, our lung capacity erodes, and our joints become more arthritic.

Now here is a tricky issue. How do we know that our rationale is correct? We don't. That is why it is a research question (as opposed to a research answer). A research question is a hypothesis; it is a hunch about how the world works. We begin by postulating how the variables in the research question are related and go on to conduct a study to assess whether our claim is true. We gather evidence to test our research question. This is tricky only in the sense of understanding how the research process works. We think our rationale makes sense, but the research we undertake is designed to test that claim. Figure 4.4 illustrates these first six steps.

FIGURE 4.4 AN ILLUSTRATION OF STEPS 1–6

Step 1: Stating the research question

Does television viewing distort people's estimates of affluence in the real world?

Step 2: Identifying the variables in the research question

Variable i: amount of television viewing

Variable ii: estimates of affluence

Step 3: Defining variables

Variable i: amount of television viewing—number of hours of TV viewing per week

Variable ii: estimates of affluence—estimates of the percentage of families owning their own home, swimming pool, luxury car, wine collection

Step 4: Specifying the independent and dependent variables

Dependent variable is "estimates of affluence"

Independent variable is "amount of TV watched"

Amount of TV watched ————→ Estimates of affluence

Step 5: Thinking causally

Is it reasonable to assume the relationship between the independent and dependent variables meets the following three conditions?

i. Is it plausible to think TV viewing and perceptions of affluence are related?
ii. Is it plausible to presume that TV viewing occurs prior to our perceptions of worldly affluence?
iii. Are there other variables that might cause both TV viewing and perceptions of affluence?

Step 6: Embedding a rationale in causal thinking

Why do we think that the relationship between TV viewing and estimates of affluence exists?

Rationale: TV programming, especially game shows and soap operas, depicts a world of wealth and fortune. Therefore, the more people watch TV, the more likely they are to believe higher percentages of families enjoy the affluence of a private home, luxury cars, swimming pools, and wine collections.

Source: Adapted from T.C. O'Guinn and L.J. Shrum (1997), The role of television in the construction of consumer reality, *Journal of Consumer Research*, 23, 278–94.

FIGURE 4.5 AN ILLUSTRATION OF STEPS 7–10

Step 7: Relating the research question to the literature

Read articles relevant to research question; two suggestions:

O'Guinn, T.C., and Shrum, L.J. (1997). The role of television in the construction of consumer reality. *Journal of Consumer Research*, 23, 278–94.

Luigino, B., and L. Stanca. (2006). Income aspirations, television and happiness: Evidence from the World Values Survey. *Kyklos: International Review for Social Sciences, 59*(2), 209–28.

Step 8: Introducing other variables

Type of program watched. Why? Watching soaps is more likely to influence estimates than is watching nature programs.

Level of education. Why? Individuals with higher levels of education typically watch less TV. They may also have a broader worldview and a more accurate conception of prosperity and poverty.

Step 9: Revisiting causal connections

A1. "Level of education" influences the "amount of TV viewing," "type of programming watched," and "estimates of affluence."

A2. "Amount of TV viewing" influences "type of programming watched" and "estimates of affluence."

A3. "Type of programming watched" influences "estimates of affluence."

B. Review the links between all variables.

Step 10: Drawing a causal model

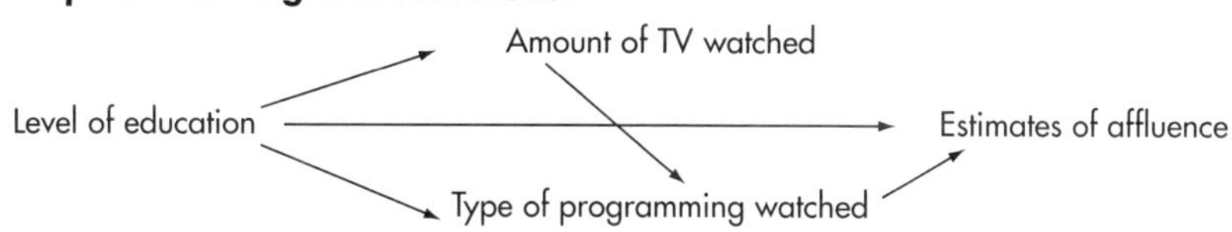

Source: Adapted from T.C. O'Guinn and L.J. Shrum (1997), The role of television in the construction of consumer reality, *Journal of Consumer Research,* 23, 278–94.

STEP 7: RELATING THE RESEARCH QUESTION TO THE LITERATURE

The research literature, which can be an important source of information and ideas for developing your research questions, includes professional journals, periodical trade publications, government publications, news magazines, and books. The literature may be argumentative and lack supporting evidence of a systematic nature; it may be ideological in approach or merely repeat folklore—hence the need for carefully evaluating it.

Think first, then read. After posing an initial statement of the research question, the researcher needs to spend some time using her or his ideas and experience to clarify exactly what is at issue. Many people immediately head for the library rather than trying to zero in on a problem for themselves. If the ideas are too fuzzy when the reading begins, the research literature can be overwhelming. Frequently, the consequence is the formulation of a new problem that is essentially defined by someone else.

In cases where there is little available literature that is related to the research question as it has been specified, the researcher needs to examine the level of abstraction to see if there is a relevant body of literature to which the problem can be generalized. For example, a research problem that involves conflict between residents within a college dormitory—a topic with a deficient literature base—may be examined from two perspectives: one that involves looking at the problem of social conflict in general, and one that looks at the processes controlling small groups, both formally and informally. The literature on both of these reformulated perspectives is quite extensive.

STEP 8: INTRODUCING OTHER VARIABLES

Everything in the world is connected. A volcano erupting in Hawaii affects weather in North America, while warm currents in the Pacific cause flooding in the Maritimes. Connections between variables in both the natural and social worlds are commonplace. In your research question, other variables will undoubtedly be connected to the variables you have included.

Consider the following research question: After you have completed school, does your level of schooling influence your employment earnings? While there is good evidence to support this claim, it is also the case that other factors influence both schooling and earnings. For example, it is not just schooling levels but fields of study that influence how much you earn. Additionally, your parents' socioeconomic status influences how far you go in school, and your parents' connections or networks also influence your prospects for employment and earnings. So now we have four variables: level of schooling, field of study, parental socioeconomic status, and employment earnings. As you think about your research question, you need to identify related variables.

There's one rule of thumb to keep in mind here. The variables of most interest are those related to both our independent and dependent variables, because these linkages help us explain how causal flows operate in our research question. In descriptive research, it is typically the case that we simply identify a longer list of independent variables that affect the dependent variable—but sometimes these variables are unrelated to one another. More interesting questions, developed when we are trying to explain or change something, involve understanding causal connections. Causal connections mean that the new variables we identify are interrelated.

STEP 9: REVISITING CAUSAL CONNECTIONS

How do these new variables relate to the independent and dependent variables? That depends on the causal connections between the new variables and our initial research question. If a variable comes between the independent and dependent variable, it is known as an *intervening variable*. To say that it is intervening is to say that causally the independent variable influences our new variable, which in turn affects the dependent variable.

If one of the new variables comes before the independent and dependent variables, it is known as an *antecedent variable*. That means that this new variable is causally prior to other variables in the research question. If a causal chain runs from A to B to C, A is an antecedent variable, B is an intervening variable, and C is the dependent variable. Notice too that in the relation between A and B, B is the dependent variable. An important message is contained in this simple illustration—the name we give a variable (i.e., independent, antecedent) depends on exactly what other variables are included. Variable B is dependent on A but is independent if we consider just B and C, and is intervening if we consider all three together.

FIGURE 4.6 DEVELOPING A CAUSAL MODEL

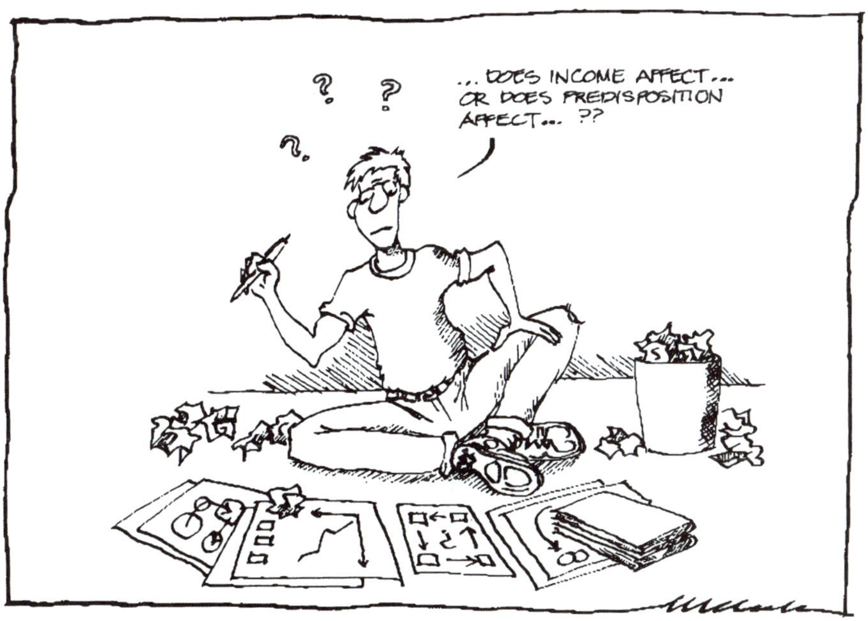

STEP 10: DRAWING A CAUSAL MODEL

After you have identified the variables of interest, clarified their meaning, and established the relationships between them, it is useful to arrange them in a diagram. Doing so helps you to visualize the argument. Good social science problems are never as simple as X causes Y. The social world is simply too complex for that. On the other hand, care must also be taken to avoid overcomplicating the research problem or defining problems that border on speculation or unsubstantiated theorizing—that is the other extreme. Generally speaking, however, good theories move beyond the realm of simply stating X and Y.

Although we started with only two variables in our research question, we have now added other variables to create a richer problem. In essence, our research question has become a research problem, with many pieces. Remember that too many pieces, or too many variables, can be problematic. To avoid an overcomplicated theory, some streamlining is essential. One method of streamlining is to organize the relationships on a large sheet of paper, with the independent variable on the left-hand side and the dependent variable on the right. The two variables are then connected, and connections are made among the intervening variables, if there are any. Finally, the antecedent variables that have been postulated are sketched in. The result is a causal model (see Figure 4.5).

In drawing your causal model, following a few simple rules ensures you will create a good visual schematic diagram of your ideas. First, put the dependent variable at the extreme right-hand side of the diagram. Do this by drawing a small box inside which you write the name of your dependent variable. Second, focusing on your research question, put the independent variable from that question several inches to the left of the dependent variable, but on the same horizontal plane (and again put this variable in a small box). Third, draw a straight line from the independent variable to the dependent variable and put an arrow on the end pointing to the dependent variable. This straight line, with its single-headed arrow, depicts the causal flow of your research question. Other straight lines with single-headed arrows in your model will represent other causal flows. Notice that, by convention, the causal flow between variables runs from left to right (see Figure 4.7).

Next, if you have any antecedent variables, place them even farther to the left, either slightly above or slightly below the independent variable named in your research question. Draw a straight line with a single-headed arrow from each antecedent variable to the independent variable of your research question. If you think your antecedent variable(s) also affects the dependent variable, draw that line as well. Fifth, place any intervening variables just to the left of the dependent variable, again a little above or below the horizontal plane. Draw

FIGURE 4.7 A CAUSAL MODEL

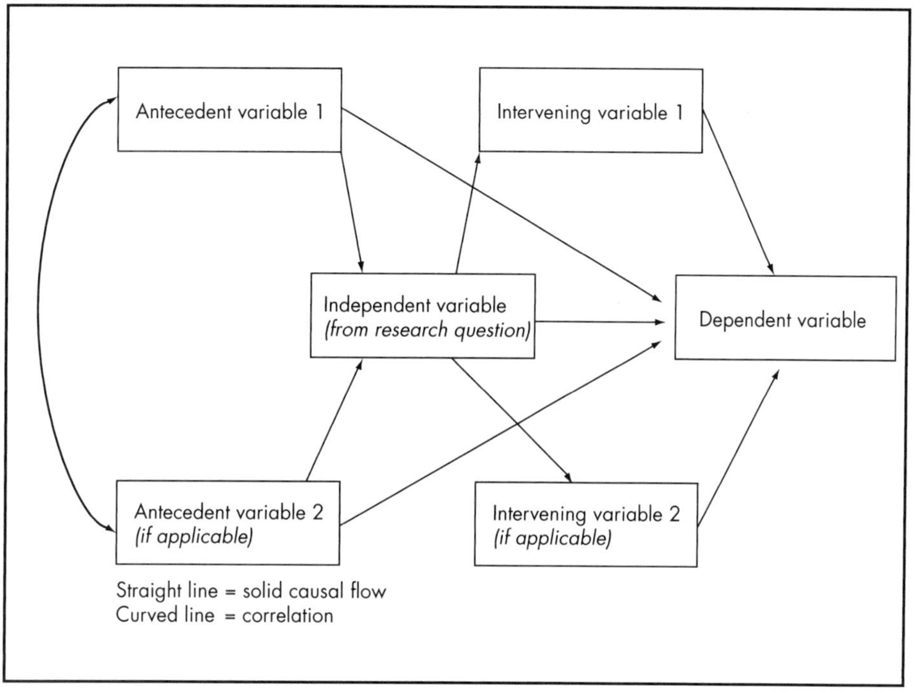

causal lines both toward and away from these intervening variables, showing how you envision the causal flows of your model. Note also that there could be causal links between individual antecedent and intervening variables.

Finally, on occasion, variables in your model will be correlated but not causally related. For example, you might be interested in how family income affects the school achievement of sons and daughters. In a causal model where each parent's education is an antecedent variable, influencing both family income and a child's education, the relation between a mother's education and a father's education is not causal but there definitely will be a correlation. To represent this correlation, researchers conventionally draw a curved line with arrows at both ends.

SUMMARY

The success of a survey research project depends heavily on the development of a good research question. This is undoubtedly the most difficult part of the research effort, as well as the most satisfying. It involves clearly identifying a research question and turning the concepts involved in the question into measurable variables. The dependent variable is defined, the independent variables are defined, and the initial causal connection between them is made. You need to stand back and think critically about the question you have posed. As well, you need to do some reading related to the question to see if you can improve it.

The paths of causal connection between the independent variable(s) and the dependent variable are then developed. Your thinking and reading will be crucial at this stage of development. Further causal connections are made by establishing intervening or antecedent variables that answer the question "How does X affect Y?" You need to ensure at this stage that your problem does not grow too unwieldy. Stick to a model that incorporates between four and six key concepts. Draw a picture, a causal model, to help you and others visualize the full research problem you have developed.

EXERCISE 4.1 IDENTIFYING VARIABLES

In Exercise 2.1, some of the statements have independent and dependent variables.

1. Identify the statements.
2. Label the independent and dependent variables.

EXERCISE 4.2 INITIAL VARIABLES

Build on Exercise 2.2 by completing the following tasks:

1. Specifying the independent variable(s) and dependent variable in your problem.
2. Finding at least three articles from professional journals that relate to your problem.
3. Providing an evaluation of those articles.
4. Specifying other variables that might affect the relationship.

EXERCISE 4.3 SPECIFYING A CAUSAL MODEL

Develop a causal model from Exercise 2.2 and/or 4.2.

1. State the variables that intervene between, or are antecedent to, the independent variables and the dependent variable. Use a schema with symbols for variables, single-headed arrows for causality, and double-headed arrows for relationships, then define the symbols.

2. Conduct a preliminary test of the clarity and plausibility of your argument by explaining your ideas to someone with good critical or analytical skills. Make notes of those parts of your argument that (a) you have to restate in other words (i.e., that are not clear), and (b) you have to introduce along with other concepts or ideas in order to convince the other person of the plausibility of your argument.
3. Rework your theory, introducing the necessary clarity and additional variables that are required for plausibility.
4. Continue to test the clarity and plausibility of your theory until you run out of friends, relatives, and classmates who will listen to your ideas, or until your ideas become clear and plausible.

FURTHER READING

Babbie, E. (2003). *The practice of social research* (10th ed.). New York: Wadsworth Publishing.

Hoover, K., and Donovan, T. (2003). *The elements of social scientific thinking* (8th ed.). New York: Wadsworth Publishing.

Salkind, N. (2006). *Exploring Research* (6th ed.). New Jersey: Pearson Prentice Hall.

Chapter 5

DESIGNING THE RESEARCH

The previous chapter explained the careful process of turning a research idea into a focused research question. At the heart of this process lie causality and the causal model. By proceeding through a series of well-defined steps, a causal model was developed. This chapter focuses on testing the causal claims of that model.

CAUSALITY: LONGITUDINAL DESIGNS

The hypothesis that X causes Y is a simple causal claim. Here are four simple rules that must be met if this X-Y relationship is causal (these are all explained more fully as the chapter proceeds).

1. *Association*. The two variables must be associated (i.e., as X changes then so too must Y).
2. *Time ordering*. The cause must precede the effect (i.e., the occurrence of X, or a change in X, must precede Y, or a change in Y).
3. *Eliminating alternatives*. To claim that X causes Y we must assume that no third variable is the real cause (i.e., it must be X and not Q or W or some other variable that causes Y; this rule is sometimes referred to as "spuriousness"—more below).
4. *Theoretical explanation*. There must be some understanding of the process by which X causes Y (i.e., we must specify some reason why X causes Y).

Figure 5.1 illustrates how we might think of the causal claim "If X occurs, then Y will occur." Here X occurs at Time 1, followed by the occurrence of Y at

FIGURE 5.1 CAUSALITY EXTENDS OVER TIME

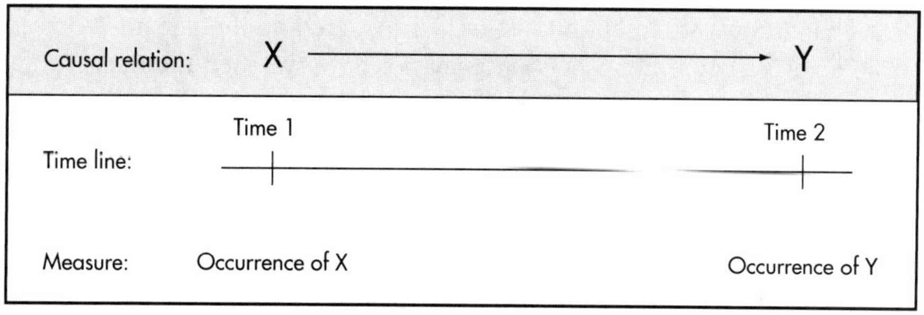

Time 2. But how do we know that the occurrence of Y is the result of X having occurred? Answering that causal question is what makes research both difficult and interesting.

Here is an example of a research design that tests causality. Televised political debates have become a central part of political campaigning. But what effect, if any, do these debates have on voting decisions? Here is a four-step process to assess the causal impact of a political debate.

1. Ask a sample of potential voters about their voting intentions (Time 1).
2. Immediately after a televised political debate, ask these same people if they watched the debate (Time 2).
3. For those who watched the debate, ask them about their voting intention (Time 3).
4. Determine how many people who watched the debate at Time 2 switched their votes between Time 1 and Time 3.

Figure 5.2 illustrates the research design discussed in steps 1 through 4, where Y_1 is the initial measure of voter intention, and Y_2 the second measure.

FIGURE 5.2 MEASURING Y BEFORE AND AFTER X

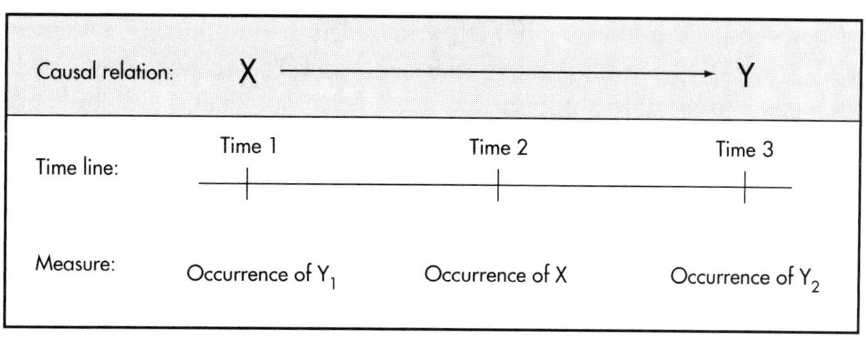

Notice, too, that we could also look at those voters we sampled at Time 1 who did not watch the debate. Their Time 3 vote intention would allow us to examine changes in vote intention not linked to watching the debate. (A complicating factor is that although people may not have watched the debate they may have heard about the candidates' performance in the debate. Our research design might minimize this complication by having Time 3 occur immediately after the debate.) The strongest research design in determining the causal effect of an event such as a political debate is to ask the same individuals about their intentions before and after the debate. This *panel study* is a strong research design because it reveals how things change over time.

Despite their acknowledged strengths, panel studies still possess certain weaknesses. First, people answering the first survey at Time 1 may not be available at Time 2 (or at any subsequent times). This may occur because respondents refuse to participate a second time, they may change addresses between survey times, they may die or become otherwise unavailable, and so forth. Lengthier time intervals between the first and subsequent surveys result in higher losses among respondents (known as sample mortality).

Second, the very fact of asking people questions at Time 1 may influence their subsequent actions. For example, if we ask people how they intend to vote at Time 1, this may trigger awareness (or greater awareness) of the political campaign. If we had not asked, their behaviour might have differed, and so it becomes difficult to generalize our results to others who were not asked about their voting intentions. This happens because our research has actually changed how people act (this is an example of Heisenberg's uncertainty principle—the very act of a single measure of a liquid's temperature changes the temperature unless by chance our thermometer is at exactly the same temperature as the liquid).

It is possible to approximate panel studies by asking people questions about previous events. These questions, known as retrospective questions, attempt to capture a respondent's state of mind or behaviour at some earlier point in time. Questionnaires often ask people to recall the education they have obtained or their date of birth. Given the salience of these events, recall is typically very good. Asking people how they intended to vote prior to a political debate is less likely to result in good information simply because many people will be unable to recall this information accurately, especially if politics is not something they follow seriously.

Designs where we take measurements at multiple times (e.g., panel studies) are referred to as *longitudinal designs*. These powerful designs help in pinpointing causal influences because we can see how variables change together over time (or do not change).

OTHER VARIABLES: MULTICAUSALITY

The notion of causation in survey research typically means X has an effect on Y, not that X completely determines Y. The idea of X affecting Y allows for the possibility that other variables will also affect Y. For example, differences between students in course performance (Y) can be affected not only by academic ability (X_1), but also by study habits (X_2) and motivation (X_3). Typically, social science research involves multiple causal variables (as reflected earlier in our causal models).

SPURIOUSNESS AND MISSPECIFICATION

Another key feature of causality highlights the correct attribution of causation. If one attributes causation falsely, or distorts the true causal flow, errors of spuriousness (falsity or distortion) and misspecification occur. This is the "eliminating alternatives" rule from our list at the beginning of the chapter.

In our earlier example, we asked whether watching the political debate had a causal effect on voting. We attributed change in voting intention to having watched the performance of the candidates in the debate. But what if, at the same time as the debate, a jogger had been the victim of a vicious attack, inciting public outrage? If the candidate receiving the greater voter support at Time 3 were the strong "law and order" candidate, it could be the attack on the jogger (X_2) and not the debate (X_1) that influenced voter change. Attributing the causal effect to X_1 (the debate) could be a case of attributing causation incorrectly (see Figure 5.3).

We have already alluded to a classic example of incorrect causal attribution earlier, in the discussion of the fire scenario (see Chapter 4). You can observe that the greater the damage done by a fire, the more fire engines are present. It

FIGURE 5.3 MISSPECIFICATION AND SPURIOUSNESS

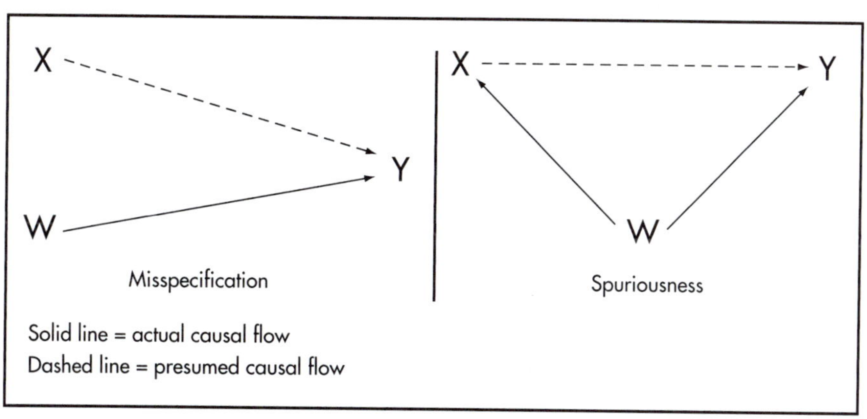

would, however, be wrong to assert that fire engines cause the amount of damage. We know this would be wrong because it is the size of the fire that causes both more fire engines to be present and more damage to be done. The size of the fire is an antecedent causal variable affecting both the response of fire engines and the amount of damage (see the right-hand panel of Figure 5.3).

What do we do to reduce the risk of improperly specifying causal connections? The key process here is to use our theoretical knowledge to incorporate properly the major plausible connections between variables. Thinking hard about causal models allows us to construct plausible models, minimizing the likelihood that key variables have been ignored. If we specify correctly our map of the variables in a causal model, we eliminate errors of spuriousness or causal misattribution.

EXPERIMENTAL DESIGN

The experiment was one of the most important innovations in science. The beauty of the experiment is that this method of research works to prevent other variables from confounding the causal link between X and Y. Experiments are effective ways of ensuring that it is change in X, and only X, that affects Y.

For social science research, the control group is of central relevance in experimentation. By comparing what happens to an experimental group and a control group, we work to isolate the effect of X on Y. Here is how the experimental design allows us to answer questions of causality.

We begin with a group of individuals and assign them randomly (by using some model of chance, like coin flipping) to either the experimental group or the control group. While the dependent variable will be the same in both groups, the independent variable, X, will be activated only for the experimental group. After assigning people to groups we take a first measurement of our dependent variable, Y. Next we expose the experimental group to our independent variable, X (e.g., a political debate, a film, a lecture on nutrition). Then for both groups we re-measure the dependent variable, Y. Finally for both groups we compare their before and after scores (the subscripted Ys in Figure 5.4). Any change between the first and second measurements in the control group gives us a baseline measure against which we can contrast change in the experimental group.

By using randomization to ensure that the experimental and control groups are composed of similar types of people, we can be confident that any difference between groups in the Y_2 and Y_4 scores is a consequence of our experimental intervention (X). Be sure you understand why the similarity of the composition of the experimental and control groups is critical.

Here is an example illustrating an experimental study as applied to survey research. The research question is this: Does providing a cash incentive increase

FIGURE 5.4 EXPERIMENTAL DESIGN

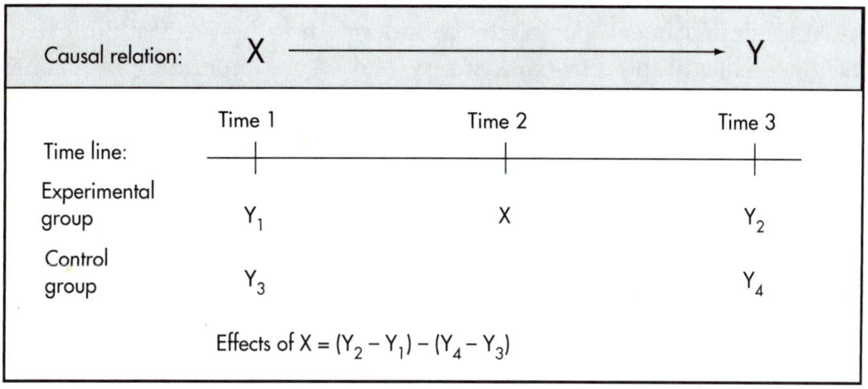

the likelihood people will respond to mailed questionnaires? Using a list from the registrar's office, we generate a representative sample of students. From this list we place every second student into an experimental group, with the others assigned to our control group. When we mail out our questionnaires, we include no cash incentive for the control group but $5 for the experimental group. Monitoring our response rate, we see whether more surveys are received from our experimental group. If there is a greater response rate from the experimental group, we can safely attribute it to the cash incentive, since this is the only systematic difference between the two groups. The cause of any increased response rate is due to the effects of the independent variable, the presence or absence of a cash incentive (see Figure 5.5).

FIGURE 5.5 EXPERIMENTAL DESIGN IN A SURVEY SETTING

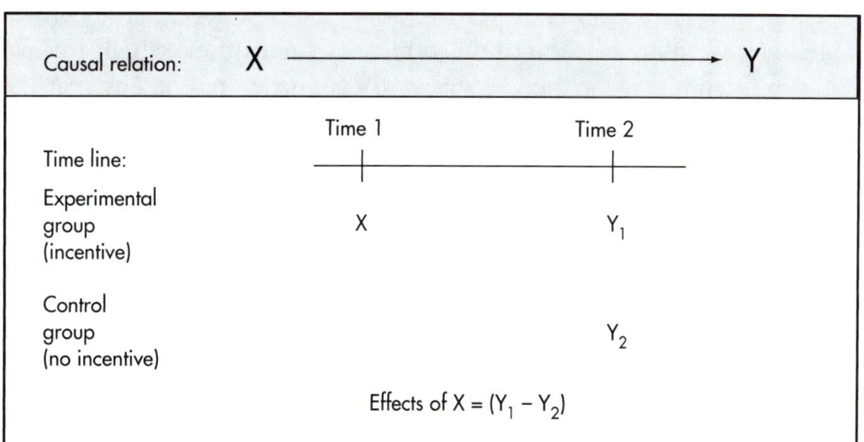

But how do we know that $5 is the magic incentive value? What if we offered only $2 or increased the amount to $10? We decide on the content (or value) of the independent variable by considering how best to test our causal claim. Knowledge of the research literature and a good grounding in theories applicable to the research question help in formulating strong research designs.

The elements of an experimental design include

1. experimental and control groups,
2. random assignment of people to experimental and control groups,
3. experimental control over the timing and content of the independent (experimental) variable, and
4. measurement(s) of the dependent variable.

Experiments isolate causal influence by eliminating the possibility of spuriousness or misspecification. As well, the researcher has control over the timing of events so that she or he can be confident that the independent variable caused the dependent variable. However, many interesting and important research questions cannot be answered using experimental designs.

CROSS-SECTIONAL DESIGNS

Most surveys use cross-sectional research designs, where respondents are asked questions at one single point in time. These designs can be likened to single snapshots from a camera, as compared to a continuous longitudinal view provided by a motion picture. With cross-sectional studies, only by asking questions of a retrospective nature can any sense of history be added (e.g., When were you born? When did you start college or university?).

Gauging the temporal order of independent and dependent variables in a cross-sectional survey requires inference from theory and experience. For example, we can infer that education influences occupation, which in turn affects current employment income, because experience teaches us that the causal chain runs from education through occupation to income. We would also postulate that education affects income directly because for people in the same occupation, educational differences are one reason that similarly employed people earn different incomes.

Typically, survey researchers also make other assumptions about causal order. Frequently people infer that attitudes cause behaviours. For example, many people assume that for households to engage in more environmentally friendly behaviour, household members must have pro-environmental attitudes. A similar assumption underlies the edict that we must teach people about the harmful effects of smoking if we are to convince them not to smoke (or to stop smoking). The moral is to change attitudes via education in order to affect behaviour.

This causal linkage between attitudes and behaviours, however, is not quite so simple. Words and deeds do not always go together (as New Year's resolutions indicate). Furthermore, people often do things, like recycle, even though they do not have very strong attitudes or opinions about "reducing, reusing, recycling" (i.e., they recycle because of blue box programs that make such behaviour easy and because their friends and neighbours are recycling). What complicates this even further is that people frequently develop attitudes as a consequence of their behaviour. They rationalize their behaviour after the fact, which complicates causal links between attitudes and behaviour.

CAUSALITY: DIFFERENCE AND CHANGE

Attributing causality is also complicated by different types of variables. To argue that gender causes something, we do not force people to change their sex in order to establish causality. Pay equity research is based on examining differences in pay between men and women employed in comparable jobs. It is not based on inferences about earnings made after someone changes sex. Similarly, we know that people living in different regions exhibit different levels of support for Canadian nationalism. Once again, we compare differences between people in different regions as opposed to the effects of people actually moving from one region to another (these are two very different research questions).

In cross-sectional research designs, we typically infer that, for example, education affects income by comparing the earnings of people who have different levels of education. The research design involves computing the average earnings of people in different education categories (e.g., less than high-school graduate, high-school graduate, some post-secondary schooling, post-secondary graduate). This is a very different research design than is a comparison of how earnings change as a consequence of individuals returning to school and increasing their education. This latter design requires a longitudinal study in order to assess the effects of having changed levels of education.

Finally, it is important in attributing causation to be aware of what a variable actually measures. What does it mean to claim that education affects earnings? There could be several reasons for education affecting earnings: a person's (1) cognitive skills, (2) social networks, or (3) certification. Across different levels of education, each of these three factors will be more or less important.

A similar issue is involved in thinking about how region influences nationalism or how sex affects income. A good deal of research is devoted to trying to understand the actual causal linkages involved in research questions. Why does sex influence earnings? Is it because of employer discrimination, or because women take on greater domestic responsibilities than do men? In essence, "sex" is shorthand, and researchers work at understanding what it is about sex (or region) that creates its causal impact.

ADDITIONAL CAUSAL VARIABLES

In discussions of pay equity, we often hear phrasing such as "comparing women and men in similar jobs." For example, to claim that women typically earn less than men we would like to know that this is not because women work shorter hours, take longer vacations, have lower levels of education, have less work experience, and so forth. We want to avoid spuriousness or misspecification, as discussed above.

A powerful comparison would be to contrast the earnings of women and men in identical jobs. However, the gendered segregation of work means few men are kindergarten teachers and few women are orthopaedic surgeons. Another way of comparing the earnings of men and women is to look at similar jobs (e.g., public school teachers) and adjust for differences between the sexes in age, experience, training, and so forth.

In the jargon of survey research, we need to "control for" the effects of other variables. Controlling for other variables allows us to eliminate alternative explanations for our research findings. When we examine the relationship between ability and performance among students, we would need to adjust for, or control for, differences in study habits and motivation. Survey researchers cannot use randomization to eliminate the effects of other variables. Instead, survey research designs are most powerful when all of the relevant variables are measured and then controlled in the data analysis (controlled statistically).

Scope Conditions

Many statements about the world must be of a conditional or limited kind. We know, for example, that water boils at 100°C *at sea level*. We know also that breast cancer is a leading cause of death *among women*. In each of these statements, the range of applicability of the claim is limited.

The scope or generality of research claims is also limited. The claim that religious beliefs are the most significant guide for most people's day-to-day behaviours is a false claim *in Western industrial nations*. However, this claim is more applicable to some other countries, especially countries with strong religious traditions. In those countries, religious beliefs serve as essential guides for many people in their daily lives.

Scope conditions serve to limit the range of applicability of our causal claims. Our earlier claim about televised political debates and voting is too simple a statement to be generally applicable. For some people, politics is irrelevant. Their level of alienation from the political process is such that they just don't care about candidates and voting. A more plausible claim is that for individuals who actively participate in the political process, political debates influence voting intentions. We could gauge political participation by asking people whether they read newspaper articles about politics or listen to candidates' speeches. For the group of people who do participate in the political process, we might then examine whether a televised political debate influences their voting intentions.

But even here we have a statement that is too simple, because those individuals who are staunch party supporters are very unlikely to change their vote just because of the outcome in a single debate. So an even more plausible claim is that for individuals who participate in the political process, but who are not firmly convinced of their voting intentions, political debates influence voting intentions.

Our claims about inequality in the earnings of men and women apply only to individuals in the paid labour force. This is a very obvious claim, but it is only under circumstances where both women and men are employed that pay equity claims will hold. By examining only employed members of the labour force, we know that our evidence on pay will not be distorted by including people who are unemployed or engaged in unpaid labour. Figure 5.6 lists factors to consider when establishing scope conditions.

An important caution when establishing scope conditions is to recognize why a particular causal claim is limited. Some claims are limited simply by sampling restrictions. For instance, the Toronto area has a municipal governance

FIGURE 5.6 ESTABLISHING SCOPE CONDITIONS

A scope condition can be understood as the circumstances under which a causal claim is expected to hold true (Foschi, 1997). Here are a few things to keep in mind:

- ❑ It is possible to have more than one scope condition.
- ❑ Make sure that your scope conditions are not actually independent variables.
- ❑ Irrelevant variables should not be used as scope conditions, whether they seem to be theoretically important or not.
- ❑ Distinguish between scope conditions and limitations of the study.
- ❑ Ensure that the participants in your survey fall within your scope conditions.
- ❑ Scope conditions determine what can and cannot be taken as evidence supporting or refuting your hypothesis.

structure that is unlike other jurisdictions. When evaluating policy proposals for reforming this structure, our findings would be applicable only to Toronto. This is, however, a sampling restriction and not a restriction on the scope of the causal claim. You must distinguish between scope conditions and restrictions of practicality. For example, you may conduct a study of single working mothers in Edmonton alone, while your research question may *theoretically* apply to single working mothers in industrialized countries. Geographical restrictions on sampling or time and money limitations on sampling do not constitute scope conditions. These are limitations of the study.

The central point here is to emphasize the limitations of research claims. We know the workings of the social world are complex; one way to cope with complexity is to avoid unbridled assertions of gross generality. By refining the scope of assertions about causal connections, we are recognizing that complexity.

SUMMARY

Collecting evidence to test the causal claims of our research requires careful design and planning. Longitudinal designs—of which experiments are a subset—provide strong methods for assessing causality. In examining causal claims, we must build strong designs that minimize risks of spuriousness and misspecification, which arise when we attribute causal influences incorrectly. Most surveys use cross-sectional research designs. Establishing causality is especially problematic in such designs. Constructing scope conditions is one way of ensuring that our causal claims about the world are applicable.

EXERCISE 5.1 REVIEWING YOUR CAUSAL MODEL

1. Think about alternative explanations for the causal linkages involved in your research question. Brainstorm. What antecedent and intervening variables would you need to include to capture possible alternative explanations? Think about how you could measure these variables.
2. Re-draw your causal model (from Exercise 4.3) to include any new or revised antecedent and intervening variables.

EXERCISE 5.2 ADDING SCOPE CONDITIONS

Identify scope conditions for the model you developed in Exercises 4.3 and 5.1. List those conditions that restrict the scope of the theory: i.e., to whom or to what does the theory apply?

FURTHER READING

Campbell, D.T., and Stanley, J.C. (1969). *Experimental and quasi-experimental designs for research.* Chicago, IL: Rand McNally.

Gray, D. (2004). *Doing research in the real world.* New York: Sage.

Hakim, C. (2000). *Research design: Successful designs for social and economic research* (2nd ed.). New York: Routledge.

Spector, P.E. (1981). *Research designs.* Beverly Hills, CA: Sage.

On the Internet

William M. Trochim's *Research Methods Knowledge Base*: http://www.socialresearchmethods.net/kb

Chapter 6

MEASUREMENT

Measuring is a familiar activity. Finding the dimensions of an object by using a ruler or determining the temperature of a room with the aid of a thermometer are common experiences. In both examples, an instrument (a ruler or thermometer) is used to indicate some value (number of centimetres or degrees) of a concept (distance or temperature).

In surveys, questions serve as measuring instruments. Their purpose is to provide measures for conceptual variables. In this chapter we begin by focusing on concepts, and then we ask how best we can measure these concepts, once clearly defined, using survey questions.

People differ in the amount of education they have. How could we measure "level of education"? We could define the concept of education by years of schooling, academic knowledge acquired, paper credentials, or some combination of the three. Analogous to the measurement of distance by a ruler, we need to specify exactly how conceptual variables (i.e., distance) will be measured using *empirical indicators* (i.e., centimetres).

We next examine units of measurement, the actual entity on which our measures focus. Research questions can focus on different "units," including, for example: individuals, communities, organizations, or the acts of individuals. Level of education, for example, can characterize a person, a neighbourhood, or a city. It is often assumed that surveys can be only about individuals because survey questions are asked of individual people. But people can report on different units of measurement—their household, their work setting, their social groups, or last week's shopping experiences. The researcher must be careful not to confuse the person doing the reporting with what is being reported on.

The final part of this chapter examines errors in measurement. The researcher must understand how errors can creep into the process of measurement. Error-free measurement is the ideal, but perfection in measurement is rare (if not impossible!). Therefore, it is important to understand the types of errors that can occur during the measurement process, and what their consequences might be. *Measurement error*, as explained below, can be either systematic or random.

EMPIRICAL INDICATORS

As we shift from the conceptual level to the measurement level, we move down a ladder of abstraction. We switch from the abstract to the concrete, from general ideas to specific instances. The goal is to tie our concepts to specific observations. This involves establishing precise indicators for concepts. In our earlier examples, the ruler was used as an indicator of distance, and the thermometer was used as an indicator of temperature.

Tests, essays, and assignments are traditional ways to measure student learning. The teacher can, as illustrated in Figure 6.1, treat tests, essays, and assignments as three different indicators of the abstract conceptual variable "student learning." Sometimes the process of devising indicators is referred to as the creation of operational definitions. The objective is to specify the precise operations to be used in measuring a concept.

The word *empirical* means observable or knowable. Our survey questions are designed to provide empirical indicators or measures of our conceptual variables. Just as tape measures indicate distance, questions about years of schooling and education credentials indicate level of education. Defining concepts clearly is an essential first step in the process of creating survey questions for purposes of measurement. Once you have conceptual clarity, then you need to design indicators to serve as measures.

FIGURE 6.1 MULTIPLE INDICATORS OF STUDENT LEARNING

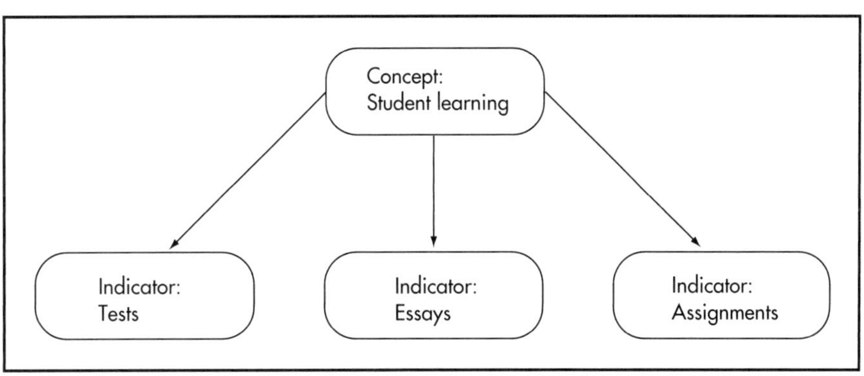

Tips for Creating Indicators

Creating a good list of possible indicators is not easy. More thought and brainstorming are required. There are a number of factors to keep in mind when generating possible indicators.

Use multiple indicators whenever possible Multiple measures are better than a single measure. Many concepts are very broad and no single indicator can reflect the full range of meaning. For example, no single question could capture what is meant by "basic arithmetic ability." A concept such as education might require multiple measures—at least years of schooling and types of certificates obtained. Some obvious concepts, such as sex, require only a single indicator: Are you male or female?

Capture the full range of a concept Indicators need to be evaluated in relation to the basic concept. Do the indicators chosen give measures of all relevant aspects of a concept? Not only do you need multiple measures of "basic arithmetic ability," but you also need to tap the full set of abilities involved—addition, subtraction, multiplication, and so forth.

Be consistent between the conceptual and measurement levels Be careful not to use indicators that are inappropriate for a concept. If social class is defined as membership in one of three discrete categories—either the working, middle, or ownership class—then earned income is not a good measure, since it is not clear what income level represents the boundary between classes. If your interest is in consumer purchases, do not use consumer preferences as indicators. A person may prefer a Porsche but purchase a Pontiac.

Ensure variability of an indicator Choose indicators that vary and that distinguish between cases. In doing survey work on environmental issues, it makes little sense to ask people general questions about concern for "a cleaner planet." On such abstract claims almost everyone agrees (i.e., there is very little variation in responses), and so the question adds nothing to what we already know. Much more variation occurs if you ask people about specific programs to clean up the environment.

Two general strategies can be used in developing a list of indicators. First, the researcher could review previous studies for appropriate indicators. Making use of other people's good work saves time and allows comparisons between the study under consideration and previous studies. Avoid reinventing the wheel. Sometimes refinements can help, but "If it ain't broke, don't fix it." The replication of findings helps researchers build a better knowledge base, but true replication requires duplicating original questions. Most good survey research studies reported in the literature will specify the precise questions used, so do some library research to help you develop your questions.

The second strategy involves talking with other informed people to help build indicators. This includes using expert or insider knowledge, reading about and interacting with target groups, and interviewing key informants—both experts and insiders. For example, a researcher asked to survey health-care workers about sterile procedures could acquire knowledge by visiting work sites, talking with workers, interviewing experts, and examining policy manuals.

UNITS OF MEASUREMENT

As noted in earlier chapters, surveys can focus on individuals or on organizations. Organizational units such as hospital maternity wards, colleges and universities, or city daycare facilities can serve as the units of measurement.

Do not confuse the reporter with what is being reported. Whether households or departments are the object of study, the information is still obtained from an individual, who is essentially an informant about that particular unit.

The "unit of measurement" refers to the specific *cases* under examination. That is, the cases could be the acts, the people, or the groups about which information is being collected. Figure 6.2 contains some examples of survey questions and their corresponding units of measurement. Note that in the last example appearing in this figure, activities defined as housework are the units being measured. The researcher is interested in gathering information on each housework activity (who carries it out, how much time is spent, etc.). This type of focus on activities is seen in survey research problems related to leisure activities, participation in physical fitness, occurrence of sexual abuse, doing homework, and visiting. The unit being measured is an activity or episode. The

FIGURE 6.2 UNITS OF MEASUREMENT

Research Question	Unit of Measurement
What factors affect after-tax profits of corporations?	Incorporated businesses
Does political alienation affect the likelihood of voting?	Individuals
Is there a link between household size and dwelling type?	Households
Do women and men share household duties equally?	Housework activities

research question relates to the properties of each episode (e.g., who does it, for how long is it done, with whom is it done, etc.).

Units of Measurement and Units of Analysis

It is useful to distinguish between units of measurement and units of analysis. For example, in a study on worker satisfaction, the focus can be on individual workers and the factors that influence their *personal* satisfaction. Alternatively, a study could be on workplace satisfaction and the variables that influence the degree of *aggregate* satisfaction among different work groups. In each instance, measurements of satisfaction would be taken on individuals (units of measurement) and expressed both as individual units of analysis (personal level of satisfaction) and as a group measure, such as an average level of satisfaction among all work group members (aggregate level of satisfaction).

Often, the units of measurement and units of analysis are the same. When these are different, the units of measurement are often aggregated and expressed as a sum, total, average, mode, rate, or other statistical summary of lower level measures. For example, unemployment rates are an aggregation of individual labour force experiences. In the example of household division of labour, information about each housework activity is summed to the household level, either in terms of the amount of time spent doing housework by the adult males compared to adult females, or by examining the list of housework activities of the females compared to the males.

Inter-Level Analysis

A research design might incorporate two or more units (or levels) of analysis. Social scientists know, for example, that context is very important in understanding social processes or events. With survey data, context is frequently studied using inter-level analyses. For example, individual job satisfaction might be influenced by the general level of satisfaction of coworkers (individuals are typically more satisfied at work if their coworkers are also satisfied). This design would incorporate two units of analysis—individuals and their workplaces—and would require a measure of individual job satisfaction as well as a measure of the average or aggregate satisfaction of others in the workplace.

An alternative study, also focused on context, might investigate individual job satisfaction as it relates to the physical size of the workplace. In this case, the second unit of measurement would be based not on aggregate information from other individuals, but rather on an attribute of the workplace—its physical size. Individuals are described on the basis of their personal level of satisfaction as well as the characteristics of their physical surroundings.

MEASUREMENT ERROR

When measurements are taken, errors occur. Researchers must work to ensure the highest quality of measurement possible. How well have the conceptual ideas been translated into empirical indicators, and how consistent or robust are the indicators as measures? In the example of student learning, one issue is how adequately tests, essays, and assignments measure what a student has learned. Students often feel that they would have done better in a course if the evaluation procedure had been fairer. Their complaint is that the evaluation methods did not adequately capture what they learned. There was *error* in the measurement process.

Two separate issues are involved. First, how well do the indicators actually tap conceptual meanings; i.e., are the indicators relevant or valid? Second, how good are the indicators at providing the same measurements in repeated attempts; i.e., do the indicators provide consistent, reliable evidence?

Systematic Measurement Error: Validity

Valid measures are those that accurately reflect the concept they are designed to measure. To return to our earlier example, if addition, subtraction, and multiplication but not division are used as indicators of basic arithmetic ability, the measurement of basic arithmetic ability will be consistently distorted. Bias will have been introduced systematically into the measurement process. In all cases, division is not being measured as part of arithmetic ability.

Likewise, if student learning is being measured, the researcher needs to ensure that the instructor's indicators of learning reflect the course strategy and objectives. If the course objectives are to develop students' ability to think creatively or to solve problems, and the course strategy is to pose problems for students to solve, then the use of multiple-choice exams to evaluate student learning is inappropriate. Systematic measurement error would exist since scores on multiple-choice exams do not reflect the concept of student learning as defined by the teacher. These examples illustrate problems that arise if indicators do not adequately reflect concepts. Measurement is impaired because our operations do not reflect our meanings. There is a persistent biasing factor at work.

Validity and bias can be examined from another perspective. Some people favour take-home exams, while others do not. Students know their own preferences on this issue—their "true opinion," so to speak. A researcher can ask them a series of questions about take-home exams in an attempt to measure their preferences. If he or she were to tell them their "measured opinion," they could see how closely it represents their true opinion. But the researcher is not directly privy to their true sentiment, and can thus never know how well it is reflected in what is measured.

FIGURE 6.3 ASSESSING INDICATORS

The dilemma here is that the researcher can never be certain of having good measures of the concepts. How closely the indicators match the conceptual meaning can never be known. The fact that he or she must measure means, by definition, that there is a gap between concepts and indicators. The point of research is to be aware of the problem of measurement error and to do everything possible to keep it to a minimum. Much of the remainder of this book is about doing exactly that.

Random Measurement Error: Reliability

Error can also occur in an idiosyncratic manner. Think about the action of weighing an item on a balance scale. If you were to repeat the weighing three successive times on a very sensitive balance, you would find that your weighings differ. If some dust landed on the item between the first and second weighings, the observed weights would differ. If you brushed off some loose particles (including the dust) between the second and third weighing, all three observed weights would differ. Alternatively, if you placed the item on different spots on the scale each time, the weights might vary. Reliability involves the agreement between the different weighings. When we have high agreement between the measures, we have high reliability. When we have low agreement between the measures, we have low reliability.

To the extent that these errors are haphazard—either increasing or reducing the observed weight—the measurement error is random. Events that are random with respect to the measurement process cause errors that are non-systematic (i.e., without a pattern). Reliability is a measure of random measurement error. When people are responding to a survey on the Web, small interruptions or distractions might influence their answers (e.g., if the cellular phone rings, if the receipt of an e-mail message is signalled, if someone calls their name). These disruptions may or may not cause errors, but if they do, the errors are likely to be random or idiosyncratic since they will occur for different respondents at different places in the survey.

The Precision of Measurement

Error is also related to precision or accuracy in measurement. If you use a ruler to measure the dimensions of a desk, you can report the distances either crudely (so many metres long) or precisely (so many millimetres long). Your choice of measurement detail is crucial. If you choose to record the crude distance, there is no way (short of re-measuring) that you can report more precise dimensions. However, if you choose to record precise lengths, you can report the dimensions in either metres or millimetres.

In a survey, the same issue frequently arises. Should you ask people their exact year of birth, or should you ask them in which decade they were born? Should you ask for exact income or for the income group into which they fall? As a general principle, it is always advisable to ask the more precise question.

By definition, crude measurement contains error. If the lengths of desks are reported to the nearest metre, any desk that is reported as two metres long could actually be anywhere between 1500 millimetres and 2499 millimetres. Likewise, someone who reports being born in the 1990s could in the year 2010 be anywhere between 11 and 20 years old. Crude measurement is easier, but the price of simplicity is often negated by the cost of error.

Discrete and Continuous Measurement

In discussing precision, all the examples we used implied a continuum—distance, age, and income. Each of these is a *continuous* variable—that is, a variable that can take on all possible values over a given range. In contrast, variables like sex, place of birth, or religious affiliation are categorical or *discrete*. Sex, having two categories, male or female, is a simple dichotomy. Religious affiliation is multi-categorical, since there are many different religions with which someone might be affiliated. With either a dichotomous or multi-categorical variable, there is no underlying quantitative continuum. The distinction between discrete and continuous variables influences the selection of appropriate statistical techniques to analyze data (Knoke, Bohrnstedt, and Mee, 2002). The distinction is also important to the types of questions asked on surveys.

A common error occurs in survey research when a categorical question is used to measure a continuous variable. For example, a survey on lower-back pain might ask, "Do you have trouble climbing a flight of stairs? Yes or No." This phrasing ignores the fact that people can experience different degrees of trouble or pain when climbing stairs. Similarly, it is generally wrong to ask people whether they support or oppose some proposition—say, an increased speed limit—and neglect to ask about the strength of their support or opposition.

It is sometimes claimed that it is not worth the effort to use continuous measures with multiple-response levels—a simple yes or no will do! This strategy confuses measurement with decision making. A thermostat measures temperature along a continuum, although a decision to turn a furnace on or off occurs at a predetermined point. Similarly, we can measure the degree of difficulty in climbing stairs and use that information to make decisions about back pain (in conjunction with other indicators). The point is to use questions that reflect continuous measurement whenever possible; do not opt for the lazy alternative of discrete measurement unless it is clearly appropriate (e.g., sex).

Improving Measurement: Scales and Indices

If multiple indicators are used to increase the validity and reliability in the measurement of a concept, some attention must be given to how these indicators are used. In measuring "political efficacy" (e.g., a person's sense of political influence), Blake (1985: 178) used seven indicators measured on a five-point agree/disagree scale (a Likert scale) (see Figure 6.4). Respondents who answered each question provided seven different indicators or measures of their political efficacy. The reliability of these indicators can be assessed by examining the amount of agreement among all of the items (the inter-item correlations). Several possible strategies can be followed in using these multiple indicators in the data analysis. Each question can be treated as a separate referent for the concept "political efficacy," or all seven indicators can be combined into a composite scale.

One approach in data analysis is to repeatedly examine the association between two variables, using as many different indicators as have been measured. For example, in examining the link between social class and political efficacy, the researcher could assess the association between class and each separate indicator of political efficacy. This strategy has advantages and disadvantages. What does the researcher make of results that show an association only when certain indicators of political efficacy are used and not others? Only if

PAIN MEASUREMENT AND SURVEY RESEARCH

The measurement of pain in medical research is a useful parallel to measurement in survey research. People interested in measuring pain, such as nurses or medical researchers, use a variety of measurement tools. One of the most precise is the "visual analogue thermometer" or the pain thermometer, which is shaped like a 10-cm-long ruler marked with two extremes of pain. The patient moves a marker along the "thermometer" to indicate how much pain she or he is experiencing. As with many survey research concepts, there is no objective way to assess pain, since pain is a subjective experience. The advantage of using a pain thermometer is that it is graduated and appears as a continuum to capture a large range of experiences. Moreover, the thermometer can be administered repeatedly in order to compare changes in pain levels over time. However, there are limitations to the thermometer: what happens if a patient indicates a level of pain at the furthest end of the thermometer ("10"), and then her or his pain gets worse?

The McGill Pain Questionnaire is an alternative, comprehensive questionnaire that measures many aspects of pain, not only intensity. It includes various scaled adjectives measuring the different elements or types of pain. For example, possible descriptors are "cool, cold, freezing." Another scale is "sharp, cutting, lacerating." The questionnaire is often administered before a treatment, and then again afterwards to measure the changes in pain over time.

Both the McGill Pain Questionnaire and the pain thermometer are useful conceptually in thinking about survey research. The McGill Pain Questionnaire, itself a "survey" of individuals' experiences of pain, uses multiple indicators that attempt to measure all aspects of the phenomenon of pain. Moreover, it contextualizes or embeds these experiences in the social and psychological milieu of the pain sufferer. It aims for precision by scaling the many aspects of pain on various continua. Likewise, the pain thermometer is conceptually useful as it appreciates that some things, like pain, cannot be measured in discrete categories of "pain"/"no pain," but on a continuum (Choinière and Amsel, 1996; Huskisson, 1983; and Melzack, 1983).

some consistent pattern among indicators emerges can the researcher make sense of the link between the two variables.

Alternatively, the researcher can combine the political efficacy indicators into a single scale before assessing the relationship between the variables. A

FIGURE 6.4 AN EXAMPLE OF A POLITICAL SCALE

(For each statement, please circle the category that best reflects your opinion.)	Strongly Disagree	Disagree	No Opinion	Agree	Strongly Agree
1. Sometimes politics and government seem so complicated that a person like me can't really understand what's going on.	1	2	3	4	5
2. The political parties are so big that I doubt I could influence them even if I were active in them.	1	2	3	4	5
3. I believe that I can help to change the minds of public officials.	1	2	3	4	5
4. People like me don't have any say about what the government does.	1	2	3	4	5
5. Generally, those elected to Parliament soon lose touch with people.	1	2	3	4	5
6. So many other people vote in elections that it doesn't matter much whether I vote or not.	1	2	3	4	5
7. I don't think that the government cares much what people like me think.	1	2	3	4	5

composite measure could be built for a concept by summing the values from the multiple indicators. For example, assuming all the indicators are valid and reliable, the seven items measuring political efficacy could be combined into one scale (see Figure 6.5). This is not always as simple as it might first appear, however. Note that item 3 is "reversed," in the sense that it is stated positively rather than negatively. To give it the same value as the other items, circling number 4 in item 3 has the same value as circling number 2 in the other items. This reversal is done by the computer at the analysis stage (see Chapter 13).

Adding up scale items is similar to adding up scores on tests, essays, and assignments to give an overall course grade. The advantages to this summation strategy are outlined below.

Tapping complexity Concepts often include many aspects, and multiple referents allow assessment of the full range of a concept. Single indicators capture only one aspect of a concept.

FIGURE 6.5 SUMMING THE VALUES OF A SCALE

(For each statement, please circle the category that best reflects your opinion.)	Strongly Disagree	Disagree	No Opinion	Agree	Strongly Agree
1. Sometimes politics and government seem so complicated that a person like me can't really understand what's going on.	1	(2)	3	4	5
2. The political parties are so big that I doubt I could influence them even if I were active in them.	(1)	2	3	4	5
3. I believe that I can help to change the minds of public officials.	1	2	3	(4)	5
4. People like me don't have any say about what the government does.	(1)	2	3	4	5
5. Generally, those elected to Parliament soon lose touch with people.	1	2	3	(4)	5
6. So many other people vote in elections that it doesn't matter much whether I vote or not.	1	(2)	3	4	5
7. I don't think that the government cares much what people like me think.	1	2	3	(4)	5

After reversing the coding for item 3 (see text):
Sum = 2 + 1 + 2 + 1 + 4 + 2 + 4 = 16

Random error reduction Using multiple indicators means that the random errors that are part of any single measure have a better chance of being cancelled out by random errors in other indicators.

Precision Multiple measures often allow researchers to make finer distinctions between cases because more information is available on each case.

Simpler analysis The analysis of the data is more straightforward, since multiple tests of relationships are not necessary. Scales summarize data.

Building scales is not without complication, however, and the following two issues must be confronted.

Weighting of items Should all indicators make the same contribution to the scale? For example, when measuring course grades we used three indicators: exam scores, essay grades, and assignment marks. Should these have equal value in assigning course grades, or should they carry different weights (e.g., make the exams worth less and assignments more)? Similar decisions must be made in deciding how to weight multiple items in forming a scale.

Missing data What happens when information is missing on a few but not all of the indicators? On occasion, a respondent might miss a question or refuse to answer a question on a survey. If this question were to be used as one indicator in a multiple-item scale, how should we proceed? Should we drop this participant's responses from the analysis, substitute an average value for the missing data based on other indicators, replace the missing data with a random value (i.e., add random error), or insert the average score from the sample for the missing data? Especially with small samples, one of the latter procedures is often essential.

SYSTEMATIC AND RANDOM ERROR

Figure 6.6 provides, by analogy, a way of understanding the difference between systematic and random error. The example of weighing the same item repeatedly (say, your instructor!) is useful. Your instructor weighs a certain amount (his or her true score), and to save possible embarrassment, let us represent that weight by the symbol "Δ". Now with your instructor's cooperation, let us weigh him or her several times, under four different scenarios.

In the upper left quadrant of the figure is a record of weights from a condition where both systematic and random error were low. In each of 11 weighings,

FIGURE 6.6 MEASUREMENT ERROR

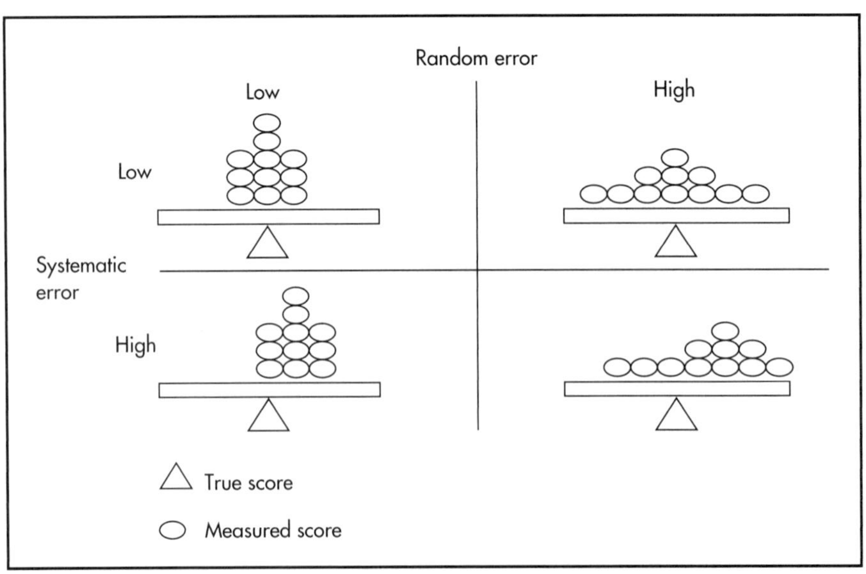

we observed and recorded weights that were nearly identical; they cluster very closely around the "Δ" symbol.

In the upper right quadrant are weight recordings under a second scenario, one in which weights vary noticeably but randomly. For instance, on different occasions the instructor held this book you are reading, exchanged the book for a piece of chalk, removed a tie or a scarf, and discarded a coffee cup. These various additions and subtractions altered the weights, but not systematically. The observed and recorded weights simply spread out more around the symbol "Δ" as the coffee cup was held and then discarded, or as the tie/scarf was worn and then taken off, and so forth.

In the lower left quadrant, the weighings show very little random error; they are bunched together closely, but they systematically overestimate the instructor's weight. This systematic error may have occurred because someone calibrated the scale incorrectly, so that it always added 3 kilograms, or the person observing the scale may have inadvertently leaned on it to keep her or his balance (thereby adding weight). Across each of the weighings in this sequence something systematically biased the observed and recorded weights.

In the lower right quadrant, both systematic and random error rates occur with greater frequency. In effect, this is a combination of the scenarios found in the lower left quadrant and the upper right quadrant. Both random and systematic error is present; as a consequence, the range of observations is wide and the observations are consistently biased upward.

The Consequences of Measurement Error

One way to think of random measurement error is to liken it to a disturbance or interference that is both chaotic and confusing. It produces haphazard or chance distortion and has three basic consequences.

First, random measurement error causes observed values to be more volatile than would otherwise be the case. For example, with more random error in the upper right quadrant of Figure 6.6, the values are spread out more than those in the upper left quadrant. Second, random error does not alter estimates of the average score. Since error occurs randomly, mistakes cancel out in the long run (or, in the example, across the 11 weighings). In the figure, the balance point is identical in the upper left and right quadrants, even though the spread (due to random error) is greater in the latter. Third, the consequence of random error is to weaken estimates of the relationships between variables. This occurs because our measurements are less consistent, and therefore the connections we estimate between relationships are less precise.

The consequences of systematic error are far more severe. Since the error is systematic, it will create bias, which results in erroneous findings. Although systematic error will not alter the spread of the observed values (so long as the amount of error is consistent), it will distort estimates such as the average score. This is seen by comparing the upper and lower quadrants on the left of Figure 6.6. In the lower quadrant, systematic error caused consistent overestimates (it could equally have been drawn as consistent underestimates). Furthermore, by consistently misrepresenting the concept we intend to measure, systematic error will lead us to biased conclusions about the relationships between variables. A linkage between an erroneously measured variable and any other variable tells us nothing of use about the actual relationship that would hold if the variables were measured accurately.

It is the goal of research to minimize both of these types of error. Indeed, the remaining chapters in this book concern techniques that are useful in obtaining high-quality measurement. One important lesson from this chapter, as outlined above, is to base measures on multiple indicators whenever possible.

SUMMARY

Good surveys demand good measurement. Before writing survey questions, it is important to clarify exactly what will (and will not) count as indicators of the concepts (as defined). Designing good indicators requires an awareness of the distinction between systematic error (validity) and random error (reliability). Finally, researchers must be clear on exactly what is to be observed—the unit of measurement.

EXERCISE 6.1 LIST CONCEPTS, DEFINE INDICATORS, SPECIFY UNITS OF MEASUREMENT

Take the list of concepts you developed in Exercise 5.1. Provide a definition for each variable (i.e., define your terms), then suggest one or more indicators for each variable. Consult other research that has measured these concepts. Evaluate these measures and develop your own. For some variables (e.g., age), this will be easy (e.g., year of birth)—less so for other variables (e.g., job satisfaction). List only the indicators you would use. In a subsequent exercise, these indicators will form the basis for questions. Finally, specify the unit of measurement for your research question (e.g., individuals, some group, or some combination).

FURTHER READING

Bearden, W.O., Netemeyer, R.G., and Mobley, M.F. (1999). *Handbook of marketing scales: Multi-item measures for marketing and consumer behavior research* (2nd ed.). Newbury Park, CA: Sage.

Rosenfeld, P., Edwards, J.E., and Thomas, M.D. (Eds.). (1993). *Improving organizational surveys.* Newbury Park, CA: Sage.

Rossi, P. (1988). On sociological data. In Neil Smelser (Ed.), *Handbook of sociology,* 131–54. Newbury Park, CA: Sage.

Spaeth, J.L., and O'Rourke, D.P. (1994). Designing and implementing the National Organizational Study. *American Behavioral Scientist, 37*(7): 872–90.

Streiner, D., and Norman, G. (2003). *Health measurement scales: A practical guide to their development and use* (3rd ed.). New York: Oxford Medical Publications.

On the Internet

METRIC, the McGill Measurement Excellence and Training Resource Information Centre: http://www.measurementexperts.org/instrument/ (and search for McGill Pain Questionnaire)

Chapter 7

THE METHODOLOGY UNDERLYING SURVEYS

RESPONDENT AS REPORTER

Social scientists collect most of their research evidence either via observation or questioning. Of the two, probably most social research is gathered by asking questions of people. People are used as reporters of information and usually, although not always, they provide information about themselves.

When physicists or chemists measure some phenomenon, they typically use instruments such as electron microscopes or spectrometers. Survey researchers' use of survey questions is analogous to that of scientific instruments. Just as there are assumptions lurking in the processes of measurement in the physical sciences, so too are there assumptions involved in the processes underlying survey research.

The process of measurement used in physics is not easily followed in social research. The questions people are asked and the responses they give are filtered through their perceptions, judgments, thoughts, feelings, and behaviours. The objects of study in the physical sciences are inert or passive in a way that the people involved in social science research are not. Using people as informants, either about themselves or about something else (e.g., an organization to which they belong), involves a number of assumptions about the appropriateness of questions and the expectations of valid answers. The assumptions

FIGURE 7.1 REPORTING THE MEASUREMENT

have to do with whether survey participants understand and know the answer to the questions asked, and are willing to admit the answer, not only to themselves, but to others as well.

The Survey Respondent Knows What You Are Asking

Asking questions depends on shared meaning Humpty Dumpty once said to Alice that every word "means just what I choose it to mean." In the world of Humpty Dumpty, communication would be impossible—no one would understand another's private language. In the researcher's world, the research results depend on shared language. Language is the cornerstone of social research, including the survey research instrument. When asking questions, the researcher uses a word (concept) to refer to some phenomenon. The respondent reads (hears) the word and thinks about some phenomenon. The researcher hopes that the phenomenon to which both the researcher and the survey participant refer is the same, and that there is a mutual understanding.

Meanings are embedded in culture Within a culture, language and meaning are shared. When the meaning of a concept is precise, the measurement problem for the survey researcher is relatively simple. However, when the meaning of a concept is imprecise or vague, the measurement problem becomes more difficult. Since language is the central part of the measurement process in the survey, the meaning of a concept must be transformed from being vague or imprecise to being clear and meaningful to all.

Precise meanings In industrial and post-industrial societies, age is defined by the amount of time having passed since the date of birth. Elaborate rituals reinforce this definition. Since almost all births occur in a hospital, a certificate of identification is issued and centrally recorded. Age, thereafter, becomes the basis for entering school, obtaining a driver's licence, buying alcohol, voting in elections, and determining eligibility for pension plans. Measuring age is a relatively simple task. We could ask the people being studied to show us a copy of their birth certificate, or ask them "When were you born?" and then calculate their age.

In our society, other concepts also have clear meanings, such as whether people have a driver's licence or own a car. Within North American culture, there is a clearly defined process for acquiring a driver's licence, as well as clearly defined rights and responsibilities that go with having a driver's licence.

Imprecise meanings A larger number of concepts in our society have imprecise meanings, so the researcher must word questions carefully, or even explain them, in order to ensure that the researcher and survey respondent are referring to the same thing. The concept of income, for example, can include or exclude income reported or not reported for tax purposes, interest income, income from stocks and bonds, or gifts of money. It can refer to net or gross income, personal or household income, and weekly, monthly, or annual income.

Asking a question about the number of people in the household may also include or exclude people who are temporarily living away at school, or who are on a work assignment away from home, or who are not officially part of the household but who may be living there on a temporary or short-term basis. Researchers must be able to clarify any such ambiguities and ask specific questions if useful information is to be obtained.

Vague meanings An even larger number of concepts in our culture are vague or subject to different interpretations. Unless these concepts are carefully considered, researchers and survey participants may not be referring to the same phenomenon. Some of these concepts include things such as visiting, authority, satisfaction, approval, free enterprise, socialism, leisure, consumerism, and patriotism. Using terms such as these in questions can lead to confusion because people may assign different meanings to any one term.

Different meanings Some concepts have different meanings, depending on the culture or subculture. For example, a billion means a thousand millions in the United States, but a million millions in England. As another example, in most Canadian universities an annual or biannual publication listing courses and programs is known as a calendar; in U.S. universities, a similar publication is called a catalogue. Thus, a Canadian who requests a calendar from a U.S. university would likely receive information showing when the school year starts, when it ends, and when breaks and holidays occur.

Similar problems within a culture exist between different subcultural or social groups who may attribute different meanings to the same word. Adults, for example, often show their ignorance by using inappropriate language when they attempt to be "cool," "hip," or "sweet," or whatever the latest slang is that defines the "right" attitude or value when interacting with younger generations.

Salient to the multicultural composition of Canadian society, cultural sensitivity must be central when designing questions and response categories. Many words are loaded politically or culturally. Terms such as *separation* or *unity* may mean distinctly different things for a Quebec sovereignist than for a Canadian federalist, though each term may have the same technical or dictionary definition. Ethnic "labels" are also significant. For example, when asking "What is your ethnic background?" response categories that are appropriate and politically and culturally sensitive should be provided (e.g., "First Nations" rather than "Indian," "East Asian" rather than "Oriental"). Survey researchers need to be sensitive to the diversity of language and cultural meanings held within the survey population.

The Survey Respondent Knows the Answer

The researcher assumes that the survey participant can answer the question. However, people are not always aware of their own perceptions, judgments, thoughts, feelings, and behaviours. Experiments in social psychology show that most of us judge attractive people as being more trustworthy, intelligent, and personable than unattractive people, but we are not conscious of doing so because of the social dictum that beauty is only skin deep.

Other experiments in social psychology show that most people read selectively, accepting evidence validating their opinions—to the exclusion of all else—and that they are not aware of this tendency in themselves. Furthermore, people are not always aware that there is often a discrepancy between what they believe and what they should do (e.g., believing that suntanning is healthy despite evidence it causes skin cancer). If the researcher simply asks people about their attitudes, beliefs, and behaviours, without being cognizant of these discrepancies, he or she is more likely to measure what people are expected to believe, not what they do believe.

People also state opinions on topics about which they know little or nothing. Not wanting to appear ignorant, survey participants might give an opinion on whether the Canadian government should grant export credits for

> **USING CULTURALLY SENSITIVE MEASURES**
>
> Zhan et al.'s (1998) study of health-promoting strategies among elderly women of different ethnic backgrounds illustrates some of these concerns of "meaning." Interviews and questionnaires were used to find out the different meanings and strategies that African-American, Chinese-American, and European-American elderly women use to maintain and improve their own health. The meanings of the strategies (which included tactics such as "getting moving," "regular checkups," "know-how," "eating right," and "keeping busy") were slightly different for the different ethnic groups. The women's measured views of medical insurance, safety of their living environment, social support, and spirituality differed and were demographically and culturally specific. Because they used culturally sensitive measures, the authors learned that many Chinese women are not encouraged to share their feelings with the family and that mental illness is a disgrace for them. Therefore, the authors warned, many Chinese-American elderly women suffer from depression and go undiagnosed and unassisted; the authors recommended depression screening for Chinese-American women.

grain shipments to the Antarctic. Questions on surveys may create opinions rather than measure them. In a classic Canadian example, Peter Pineo and John Porter (1967) asked people to provide occupational prestige ratings for a set of job titles. They included two nonexistent jobs—"archaeopotrist" and "biologer"—that were nevertheless rated by the majority of respondents.

The Survey Respondent Is Willing to Accept the Answer

Most people have had experience with others who cannot seem to recognize certain things about themselves, and who are thus said to have blind spots. The researcher has to be careful that she or he is not asking questions that set out to measure something the survey respondent is unwilling to answer or that the respondent is unwilling to accept about himself or herself. If survey participants are not likely to accept it, they cannot report it. Child abusers may define themselves as disciplinarians. Alcoholics typically deny having a problem—they "can quit anytime."

Not all self-deception is so dramatic. Minor distortions also frequently occur in what we are willing to admit to ourselves. As social animals, we subscribe to norms (standards) that are considered desirable and by which we judge ourselves and others. It is normal to gloss over our deviations from

FIGURE 7.2 SELF-DECEPTION

societal norms. Indeed, we may actually think that we are younger/older, heavier/lighter, or taller/shorter than we actually are. At younger ages, we round up our age, only to do the reverse as we get older. Weight varies even more than age: "I weigh a little more than 90 kilograms, but refuse to stand on a scale to find out that I actually weigh 96 kilograms." As researchers, we have to remember that social values and norms may determine the answers provided by survey participants. We need to be particularly aware of how social norms might affect the answers we get.

The Survey Respondent Is Willing to Admit the Answer to Others

The same process of self-deception may also operate when reporting information to a researcher. People are concerned with presenting themselves in a favourable light, and thus they tend to screen or even manipulate the information they present. Students do not admit to an instructor that a research paper will be late because they left it to the last minute. Instead, papers are late for more acceptable reasons, including the following:

> "Having to work part-time and a heavy course load made it difficult to find time to work on your paper." Similarly, people who are asked what type of work they do may say that they are in public relations rather than say they work at a tourist information booth.

The same type of "impression management" may also be operating when people fill out a questionnaire. If the questions are general or vague, respondents have the latitude to provide answers that may be literally true without being accurate representations of reality. If the question "In what type of job are you employed?" is asked of the person working at the tourist information booth, the answer "Public relations" is literally true, but it does not accurately represent the job; it gives the impression that the respondent works in an office as an executive or professional rather than as a clerical receptionist.

WHEN ARE SURVEY RESPONDENTS NOT UNBIASED REPORTERS?

For each variable in our research, our goal is to create a survey question (or set of questions) measuring a specific phenomenon while minimizing the opportunity for the respondent to engage in intentional or unintentional deception. How do we know we have a problem when asking a survey participant to accurately report on his or her attitudes, opinions, or behaviour? When there is a norm, social rule, or value that favours one response over another, respondents are more likely to engage in self-deception and/or deception of others.

How do we avoid measurement errors caused by intentional or unintentional deception? The best way is to avoid general questions that do not have a specific referent. General questions about traditional versus non-traditional divisions of labour in households are too vague. If we ask more specific questions about a set of behaviours—such as who cleared away the dishes from the previous evening's meal—the respondent is less likely to distort the report on the behaviour. The judgment regarding how this information is interpreted with respect to the household division of labour is moved from the respondent to the researcher.

Furthermore, it is imperative to pre-test the questions to be used on a survey. Even if the researcher is asking survey questions that measure the variables of interest, he or she will need a process for testing and improving the questions. This is necessary to ensure that questions are understood by survey respondents in the manner intended by the researcher. Testing questions with a small, representative group of potential survey participants is essential, since it allows the researcher to make changes that will enhance the utility of the questions asked of respondents (see Presser et al., 2004, for more detail).

ALTERNATIVES TO ASKING QUESTIONS

If asking questions of the respondent is not an appropriate tool for measuring variables, the researcher needs to look at other methods of getting the information needed to answer the research question. She or he needs to evaluate whether the information can be obtained through observation—either in natural settings or by experimentation—or by using administrative data or other types of records.

Observation

When we are concerned with measuring behaviour that the respondent is either not aware of or is likely to distort in reporting, alternatives to doing a survey need to be examined. If the behaviour occurs in situations that are normally not hidden from others, the researcher should consider using observation (e.g., participant observation). Observational studies of micro-behaviour are often most effective when only a small number of participants are involved.

In natural settings Watching behaviour in everyday settings is referred to as observation that is carried out in "natural environments." The researcher does not manipulate the setting, the events, or the behaviours, but instead observes events as they unfold and often supplements his or her observations by listening to conversations and asking questions. For example, much of the research on small-group structure, behaviour, and process, as well as body language and spacing behaviour, has been developed through making field observations. Most of the research literature on urban neighbourhoods and community studies has come about from observational work that has been supplemented by interviews. Such studies can provide extremely rich, textured material that captures important nuances and complexities virtually unattainable in surveys.

But observations in natural settings have limitations as well. Whereas some types of data gained from surveys may suffer from the bias of respondents, the data collected from observations are likely to suffer from the bias of the observers. Even trained observers who are aware of this bias often see what they want to see, matching their conclusions to their hypotheses and ignoring contrary results. Many observational studies also suffer from a lack of explicitness in the research tools: theories are not clearly developed beforehand but are constructed after the data are collected, measures are not precisely defined, sampling techniques are not developed, and coding and analysis of the data are not defined. Finally, on some occasions, the presence of the observer can cause people to act differently.

Observational studies nonetheless can contribute to knowledge. They enable the researcher to observe interactions between retailers and customers, campaign managers and potential voters, employers and employees, hospital staff and patients, and so forth. Often, however, the knowledge gained from observation needs to be systematically developed and tested under more tightly controlled conditions.

In experiments Observation under experimental conditions, another research tool available for theory testing, compensates for many of the disadvantages of observation in natural settings. Methodological explicitness, which is lacking in the literature that is based on observations in natural settings, is very much a part of the research literature based on experiments. The researcher exercises control over the variables and the testing of the hypotheses (including the temporal sequencing of the variables in testing causality, and the control of other variables that might affect the results). All these methodological components of the research become part of the research report.

Marketing research has made extensive use of experiments. The study of responses to television commercials includes experiments involving the filming of people's eye movements as they react to visual presentations. Other marketing experiments have focused on packaging materials, location of displays, and audio advertising (including experiments on background music in stores, and how it can help to sell products). Experiments are also very much a part of the human relations and organizational behaviour fields in work organizations. Experiments on communication, types of work groups, work conditions, the structure of decision making, employee training, and motivational behaviour are just a few of the areas in which experiments add to the knowledge about work and work organizations.

Experiments can be carried out in the natural setting as well as in the laboratory. A major disadvantage of this practice is that retailers, recreation administrators, and organizational managers carry out experiments in the natural setting (i.e., in their store, community centre, or firm) without adequate controls over all the experimental conditions. The knowledge gained from these experimental efforts may be considerably restricted when the limits are not known by the user of the data. On the other hand, experiments in the laboratory, although well controlled, may not be so easily generalizable to the natural setting, where the conditions are very different from those contrived in the lab.

There are other drawbacks to using experimental data for solving problems. In studies involving marketing, voting, community action, and organizational and recreational behaviour, the participants often vary, and getting similar types of people to participate in experiments is problematic. As well, the number of variables involved in measuring the diverse dimensions of the heterogeneity of people requires large experimental samples, which make the costs of such experiments prohibitive.

The Use of Secondary Data

There are many types of research problems for which data already exist, or for which data are being constantly gathered. These can include data from previous surveys or data that are gathered and recorded for other purposes. For example, retailers attempting to determine the characteristics of household members who reside near their store do not require a survey but can instead refer to federal census data, as can political candidates who need to know the characteristics of the people in their riding. Similarly, a voluntary organization interested in setting up daycare facilities for single-parent, low-income households can use census data to identify where these facilities should be located. As well, a manager who needs to know whether employee salaries relate more to length of service than to performance can consult administrative data rather than undertake a survey.

If a research question can be examined using administrative data, the researcher needs to carefully consider the conditions that dictated the gathering of data. Official statistics from administrative records contain their own problems. How is each item defined? Does the information that is filled in for each item clearly reflect what it measures? Are the items an accurate recording of the information? Is the information complete? Is there information on everyone who is supposed to be in the database, or are some records missing? Are there possible biases?

SUMMARY

After the researcher has developed his or her research question, a decision needs to be made about whether a survey can provide accurate measures of the variables. The variables to be measured need to be evaluated against the assumptions that underlie asking people to report on themselves or others. Do the respondents understand the question? Do they know the answer to the question? Will they be willing to admit the answer to themselves? Will they be willing to admit the answer to others (i.e., the researcher)?

The researcher also needs to ascertain if the data are available in administrative records or from other surveys, to evaluate observational techniques in order to determine if these techniques could provide better measures, and to decide if an experimental approach would be more appropriate to the measurement of the research variables. If the answers to these questions are negative, the researcher may conclude that a survey would be a more appropriate research tool for her or his purpose.

EXERCISE 7.1 CHECKING YOUR VARIABLES AGAINST SURVEY ASSUMPTIONS

Begin with a blank page turned lengthwise and draw six columns (with the three final columns each about three centimetres wide).

1. List all the variables in your theory in column 1.
2. For each variable in the list, specify whether you intend to measure a subjective condition (perception, opinion, belief, norm, value, attitude) or an attribute or behaviour of the respondent(s) (column 2).

3. Specify whether there might be a social norm or value that would favour some values of the variable over others (column 3).
4. Check your variables against each of the assumptions. Will the respondent know the answer (column 4a)? Will the respondent admit the answer (column 4b)? Will the respondent admit it to others (column 4c)?
5. If the answer to any of 4 above is negative, how do you propose to get valid measures?

FURTHER READING

Birn, R., and Hague, P.N. (2002). *The international handbook of market research techniques* (2nd ed.). London: Kogan Page.

Dutka, S., and Frankel, L.R. (1993). Measurement error in organizational surveys. *American Behavioral Scientist, 36*(4), 472–84.

Foddy, W. (1993). *Constructing questions for interviews and questionnaires.* Cambridge: Cambridge University Press.

Presser, S., Rothgeb, J.M., Couper, M.P., Lessler, J.T., Martin, E., Martin, J., and Singer, E. (2004). *Methods for testing and evaluating survey questionnaires.* New York: John Wiley and Sons.

Richins, M.L. (1997). Measuring emotions in the consumption experience. *Journal of Consumer Research, 24*(2), 127–46.

Chapter 8

DESIGNING QUESTIONS

Well-designed and effective questioning lies at the core of any good survey. Preceding chapters have discussed what to ask, why to ask it, and the ethics of asking it. Now the focus shifts to *how* to ask questions—what form questions should take, what wording is appropriate, and what the format of the responses should be.

OPEN AND CLOSED QUESTIONS

Two formats can be used in asking survey questions. One approach is to ask a question and then allow the respondent to provide the wording of the reply. These are called *open-ended questions* because respondents choose how to answer in their own words, analogous to short-answer questions on exams. Responses can range from a single-word to a more lengthy statement.

A second strategy is to ask a question and then provide the respondent with a list of answers from which she or he can make a selection, much like on a multiple-choice exam. These are known as *close-ended questions* because of the limited set of possible answers.

A third option, which can be used only in an interview setting, combines these strategies by asking a question without showing respondents answer categories. Instead, interviewers are instructed to classify respondent answers according to one of a series of predetermined categories. These are referred to as *field-coded questions* because interviewers code responses "in the field."

Open-Ended Questions

The following is an example of an open-ended question, often asked in national surveys such as the Canadian Gallup poll.

Question: What is the most important problem facing the country today?

How would you respond? Take a second and consider what answer you might give. Here are some typical responses:

Responses: Unemployment—too many people are out of work.

Well, taxes are pretty high.

The environment—we are ruining the planet.

The government is going bankrupt trying to clean up our dreadfully polluted environment.

The world's population is out of control.

My grandma thinks families need more discipline.

The first three responses are helpful in identifying what respondents see as important problems: they are specific, clear answers to the question. However, the last three responses are all ambiguous. Is government spending more important than environmental pollution? Is the world's population a problem for Canada too? Does the respondent agree with Grandma?

If these ambiguous responses were given on self-administered questionnaires, no further information could be gained, which would result in unusable responses. However, if an interviewer were asking the questions, follow-up questions could clarify the answer from the latter three respondents. For this reason, open-ended questions are usually more appropriate for interviews than for self-administered questionnaires.

The advantages of open-ended questions are that they

1. *provide full expression.* By eliciting details and opinions that are important to the respondents, they allow people to develop their answers.
2. *allow for the drawing of salient distinctions.* Respondents use their own words and choose what to stress and what to ignore.
3. *tap unanticipated answers.* They ensure that the researcher's assumptions do not inhibit the answers.
4. *add to the respondents' enjoyment.* So long as there are not too many questions, respondents often enjoy the opportunity to fully express their views.
5. *provide rich vignettes.* Research reports can be considerably improved by quoting the words of lucid or candid respondents.
6. *are a good first step.* For the researcher who is planning a survey in an unfamiliar area, open-ended questions provide a very good way of learning the language respondents use, and the categories they employ.

7. *work well with endless lists.* A manageable list of response alternatives is not always possible. Asking for a person's occupation is best done with open-ended questions since the list of possibilities is extremely long. Open-ended questions can also be used to measure precise income and age (e.g., In which year were you born?), as opposed to having respondents check off close-ended categories that represent the income group or age cohort to which they belong.

There are also disadvantages associated with the use of open-ended questions. These include

1. *comparability.* People may give responses that differ widely and thus are not comparable (as in the example of problems facing Canada). The result may be that the researcher receives almost as many different responses as there are survey participants.
2. *vagueness.* People may give answers that are too vague, that do not fully answer the question, or that deal with only one dimension of an issue.
3. *recording.* People are usually loath to write long responses to open-ended questions. With interviews, responses must either be tape recorded or detailed notes must be taken. In some cases, people will not write anything, will refuse to be taped, or will be put off by intensive note taking.
4. *coding/summarizing.* Verbatim reports of all respondents' answers cannot be included in research reports, and so summaries of responses must be made. But summarizing people's meanings and intentions is never easy and can lead to distortion of their answers.
5. *ordering/intensity.* Comparing the strength of respondent attitudes, or deciding who does more or less of something, is difficult. Ranking or ordering respondents along any continuum is hard to do with free-form responses.
6. *greater demand on respondents.* Open-ended questions require more involvement and time from the respondent. If the respondent is not sufficiently motivated, he or she may give only partial answers to save time.
7. *potentially greater costs.* When the sample size is beyond 200, listing all of the answers, and then trying to create a single variable from all of the answers, becomes a costly process in terms of both time and money.

Close-Ended Questions

Below is an example of a close-ended question, using the same question as before, but now with fixed response categories.

Question: What is the most important problem facing the country today?

Response choices:

_______	Crime	_______	Political leadership
_______	Debt/Deficit	_______	Taxes
_______	Education	_______	Trade
_______	Environment/ Pollution	_______	Unemployment/Jobs
		_______	Other (Please specify)
_______	Health care		
_______	Immigration		
_______	Inflation	______________	
_______	National unity	_______	Don't Know

How would you answer the question this time? Would your own answers to the open and closed questions have been the same? Research evidence suggests that the responses of people vary depending on whether a close-ended or an open-ended question is asked (Bradburn and Sudman, 1979: 14–25). Thus, the problem a respondent selects as being the most important one facing Canada would depend on the question format used. Obviously, great care must be taken both in asking questions and, with close-ended questions, in providing appropriate response alternatives.

Two key principles guide the development of checklists for close-ended questions. First, the response categories must be *virtually exhaustive* in covering all the major responses expected. Second, the categories must be *mutually exclusive*, so that respondents do not mark two or more responses. In deciding on the number of categories, remember that it is generally better to use narrower categories (which can be combined later if necessary) than it is to use only a few crude or broad categories. Also, although an "other, please specify" category can be used, relatively few respondents should feel compelled to use this alternative (otherwise, you really have an open-ended question).

The advantages of close-ended questions are that

1. *alternatives are considered by the respondent.* Because they see a range of alternatives, respondents are reminded of all possible answers. A good list will cover the full range of responses.
2. *responses are uniform.* Provided that respondents understand questions and answer choices in the same way, responses can be compared because everyone uses a common frame of reference. This is the power of systematic comparison, which we have stressed earlier.

3. *less demand is placed on respondents.* Respondents need not search through their memories to consider all possible answers, and they are less likely to respond with the first thing that enters their minds.
4. *respondents make their own judgments.* Respondents choose for themselves which of the response alternatives they prefer. There is no necessity for the researcher to make judgments on their behalf in order to summarize the data.
5. *recording is simplified.* The responses are much easier to computerize since numbers can be assigned to represent each answer category even before the data have been collected. Less recording error occurs.

The disadvantages of close-ended questions include

1. *inadequate response categories.* The list of alternative responses may not adequately capture respondents' views. The response categories may be too general for their taste, or the respondents may be unable to choose from among categories (or even find their preferred response listed).
2. *superficiality of responses.* Respondents may feel that a single response neither does justice to the complexity of an issue nor adequately captures their own views.
3. *tedium of going through long lists of responses.* Just as long lists of multiple-choice exam questions can be very boring, so too can lengthy lists of close-ended questions, which increase the frustration of respondents.
4. *inappropriateness of long lists of alternatives.* Only if the list of alternative responses is relatively short is this form of question appropriate. The list in the example above is probably too long (respondents may forget alternatives), especially in a telephone interview.

One very pragmatic issue to consider in close-ended questions is whether to ask respondents to choose only one of several closed alternatives or to check "all that apply." Most evidence suggests that the latter should be avoided because respondents provide richer, more accurate replies when they are forced to make choices. The act of choosing makes one think harder about the options (see Smyth et al., 2006).

Field-Coded Questions

In an interview situation, there is a third option. The same question about important problems facing Canada can be asked, but with no response alternatives given to respondents. Rather than recording verbatim responses, interviewers can be instructed to check off one of a set of response alternatives provided only to them. If a response is vague or off topic, the interviewer can ask supplementary questions to clarify which of the fixed categories is most appropriate.

Field-coded questions share most of the advantages of open-ended questions while minimizing certain disadvantages: supplementary questions can probe to avoid vagueness, the information can be summarized by the fixed categories supplied, and the response can be easily recorded. However, this technique can be used only in interview situations and, even then, only if the interviewers are well trained and motivated. Most questions on contemporary surveys are close-ended, and experienced survey researchers advise that open-ended questions be used cautiously. In our experience, open-ended questions work well for smaller, more exploratory or intensive investigations, but are less useful in large-scale surveys. Close-ended questions provide systematic information that can be compared more easily between respondents.

WORDING QUESTIONS

Beyond the different types of formats that can be used to frame questions, issues of question wording also require attention and comment.

Word Selection

By changing a single word in a question, you can significantly alter the responses people will give. The test-marketing of different question wordings is practised routinely by governments wishing to use referenda questions most favourable to their own position. For example, public support for social programs tends to be higher if words such as "poor" and "poverty" are used, as opposed to words such as "welfare" that frequently evoke images of waste and cheating (Smith, 1987).

Care should also be taken to avoid using sexist or racist language (e.g., "scientists and their wives") or making sexist or racist assumptions. For example, the statement "It is acceptable for women to hold high political office..." implicitly uses men as a standard, thereby precluding the response "It is *more preferable* for a woman than a man to hold high office" (see Eichler, 1991: 43–44).

Focus

Questions must have a single, specific focus. For example, market researchers often focus on purchasing behaviour. They want questions that focus on what people actually buy ("Which type do you buy?"), not on what they might prefer ("Which type do you prefer?").

Bad: Do you agree or disagree with the new exam regulations?

Better: The college is proposing that the final examination period be reduced to 10 consecutive days. Do you agree or disagree with this proposal?

The second question is better because it focuses more closely on a specific issue—it does not deal with all exam regulations, but just with the regulation regarding the time period. Another example of focus in a question is:

Bad: How much literature do you read?

Better: How many novels have you read in the past four weeks?

Not only may the focus of a question be vague, as in the "bad" examples cited above, but more than one question may be asked. An obvious example of a double-barrelled question is "Do you think city hall is managing money well and helping small businesses?" A more subtle example occurs in the following question:

Bad: Do you exercise regularly to avoid heart trouble? [Yes, I exercise regularly, but not to avoid heart trouble.]

FIGURE 8.1 ASKING PRECISE QUESTIONS

To avoid questions with more than one focus, use simple sentences instead of complex sentences. If the word "and" is used in a question, ensure that it does not connect two different concepts.

Brevity

Short, specific questions generally create less confusion and ambiguity. Respondents may get lost in the detail of lengthy questions, or they may focus on different aspects of the question. For example:

Bad: Could you please tell me how many children you have, and in what year each of them was born?

Better: What are the ages of your children?

Bad: Textbooks can be purchased at many different places. Thinking about your textbooks for this year, where exactly did you buy your books?

Better: Thinking about your book(s) for this course, where did you buy it (them)?

Occasionally you need to add words to a question to establish a context. If you were to ask health-care professionals about sterile techniques, the question would need some context:

Bad: Do you always use sterile techniques?

Better: In some emergencies, sterile techniques can be overly cumbersome. Have there been any occasions in the past seven days, since last Monday, that you have had to use non-sterile techniques?

Not using sterile techniques whenever possible is a cardinal sin in health care, so everyone is likely to agree strongly with the first version of the question. The second question admittedly introduces some bias, but it tries to establish a context in which people may feel they can safely report not using sterile techniques. Follow-up questions can then pin down reasons more precisely (see Sudman and Bradburn, 1982). Another example of a question with context is:

Bad: Have you ever driven a car after drinking alcohol?

Better: Sometimes people will drink alcohol—for example, at a social event—after which they drive a motor vehicle—for example, to get home after a party. Have you done this within the last 12 months?

Clarity

Asking all respondents the same question does not necessarily mean that each will hear the same question. Using short, plain words helps, but even simple words such as "wine" or "exercise" may mean different things to different people. Often you can add clarity by defining your terms as a preliminary lead-in to the question:

Bad:	Have you had any wine to drink in the past month?
Better:	Table wines, sparkling wines, champagne, sherries, brandies, and similar drinks are increasingly popular in Canada. Did you drink any of these in the past month?
Bad:	Do you know any vegans?
Better:	Vegans are people who abstain from eating meats and animal products, including items such as milk, eggs, and fish. Do you know any vegans?

Ambiguity can also be needlessly introduced by using negative phrasing, as in the following statements:

Bad:	Marijuana should not be decriminalized.
Better:	Marijuana use should remain illegal.
Bad:	Cyclists should not be disallowed from following the rules of the road.
Better:	Cyclists should follow the rules of the road.

Try to avoid using "not," since it often contributes to double negatives, which are hard to interpret. Common words help to ensure common understanding, although you must be sensitive to the levels of education and the cultural backgrounds of respondents. An important caveat is to write questions designed for your audience. For example, when surveying health-care workers or other members of specific professions that maintain a common body of technical terms, you should strive to use their language wherever appropriate.

Biased and Leading Questions

Care must be taken to word questions so that no particular response is favoured over another. Using strong words will skew responses:

Bad:	Do you agree or disagree that students who whisper in classes should be expelled from the classroom?
Better:	Do you agree or disagree that students who whisper in classes should be asked to leave the classroom?

Fewer people will agree with the first item because "expulsion" is too severe and emphatic a word. Conversely, it is difficult to disagree with a statement like "In order to save lives, more money is essential for health care." Everyone wants to help save lives, and especially if money is "essential." Try not to overemphasize or overstate issues because you may introduce bias into the measures. Beware, too, of "social desirability effects," which occur when people give the "polite" response by answering questions based on what is thought to be proper or acceptable. Respondents should be made to feel that all answers are viable (see Chapter 7).

Time Frames

Survey researchers have a vested interest in assuming that respondents have good, accurate memories. However, asking detailed retrospective questions about past events is extremely difficult because most of us (especially as we age) tend to recall only general information (Bachman and O'Malley, 1981; Canl, Marquis, and Laurent, 1977). Over time, details get forgotten or become blurred, and values and norms often affect how we choose to recall past events (e.g., the more unpopular a political figure becomes, the fewer people remember voting for him or her). Furthermore, event recall can be extremely tedious and taxing for respondents. Despite these problems, asking questions about past events is important for some survey research projects.

In helping people to report their past behaviour accurately (e.g., consumer purchases, visits to the doctor), the researcher should establish a suitable time frame. People's recall depends on the saliency of an event or activity (e.g., a wedding day, high-school graduation). Events that are unusual or costly, or have continuing consequences, are more salient, so more precise dates and frequencies of occurrence can be recollected. Open-ended questions with no time frame do not work well.

Bad: How often do you go to your dentist?

Better: In the past six months, how often have you gone to your dentist?

Bad: How often do you eat fast food?

Better: Over the last seven days, how many times have you eaten a fast-food snack or meal?

With the most salient events in people's lives, such as buying a car, researchers can generally use an open-ended question to establish a precise date. With topics that are less salient but still important (e.g., visits to the dentist or doctor), respondents are often asked to recall events within a short, delimited period (six months or less). With events that are commonplace or routine (e.g., cups of coffee consumed), a much shorter time frame is appropriate ("Thinking back to yesterday, could you tell me how many cups of coffee you drank?").

If a key survey objective is to gather event or episode history information, avoid asking respondents to summarize past events. The temptation is to assume that respondents have good memories and that they can summarize past events accurately. This is rarely the case. Rather than asking respondents to provide a summary estimate of their behaviour, it is better to ask them to recall actual events or episodes. Instead of having people give a summary estimate of their dental appointments, ask them to list their visits using the purpose of each visit as a memory cue. Returning to the example of coffee drinking during the previous day, more accurate recall will be attained by having respondents recall the approximate time of day when they had coffee and whether they had a refill (or two or three) each time. Based on the principle described in Chapter 7, if social or medical norms state that drinking more than three or four cups of coffee is bad for you, people who drink more will likely underestimate the amount they consumed.

There is an important difference between questions that ask respondents to summarize information and questions that allow the researcher to summarize information. In household surveys, it is often important to establish exactly how many people live at a particular address. In the first question below, the respondent is asked for the number of household members (her or his summary, so to speak). In the second question, the respondent is asked to list household members; this listing can then be used by the researcher to calculate how many people live in a household.

Bad: How many people live in this household?

Better: Could you please list the first names of the people living in this household?

As a supplement to the second question, additional information on each person might be requested (relation to the respondent, age, etc.). This has the added advantage of helping to clarify whether certain residents (e.g., a temporary guest, a nanny, a relative staying for a few weeks) should be considered household members. In our experience, asking people to list household members provides the better measure of household size.

Threatening Questions

Some, or even all, respondents may find certain questions threatening. Questions about cheating on examinations, cross-border shopping, or family disputes might intimidate respondents and thus negatively affect their answers. To avoid this, the researcher can soften questions by using an appropriate lead-in. Rather than directly asking "Did you vote in the last federal election?" more reliable data may be obtained if you ask "Were you able to vote in the last federal election?" People who didn't vote find it easier to answer no to the second question without feeling intimidated or put on the spot.

If questions about threatening issues must be asked, a proper context also needs to be established. Many people believe that striking children is wrong and would not admit to it if asked "Have you ever struck your child?" However, if you rephrase the question appropriately ("Given that sometimes even the calmest parent gets frustrated and strikes a child, have you ever struck your child?"), it is easier for people to admit that they have engaged in a behaviour they see as undesirable (Sudman and Bradburn, 1982).

Drug use is another topic that many people find threatening. However, if questions are embedded in the context of other questions involving cigarette smoking and alcohol consumption, the threat is eased somewhat. A similar strategy involves starting with a general topic (e.g., child discipline) and then focusing on specific punishments (e.g., hitting or scolding). Devising ways to ensure anonymity also helps in acquiring more accurate responses to threatening questions. However, the ethical issues raised in Chapter 3 must be addressed before the researcher asks questions about illegal behaviour, or other potentially threatening questions, no matter how sophisticated a strategy he or she uses to frame the questions.

USING ATTITUDE OR OPINION STATEMENTS

Surveys often include items designed to assess an individual's attitudes or opinions. Rather than asking a direct question, one strategy is to present a statement and then ask people whether they agree or disagree with it. The statement serves as an initial stimulus, and a set of response categories is used to measure reactions: "strongly agree," "agree," "disagree," and "strongly disagree." The adjectives used to describe each category can vary: "agree/disagree," "approve/disapprove," or "do support/do not support." This two-sided or bipolar response format is widely known as a *Likert scale* (Likert, 1932).

In using Likert scales, respondents are typically requested to "circle the number that most closely represents your view," as illustrated in Figure 8.2. Note that no middle category exists in the agree/disagree continuum. Opinion differs on the merit of including a middle category. If the goal is to determine how people are leaning on an issue, do not use a middle category. However, if the purpose is to assess the views of those with definite convictions, then the use of a middle category allows those who are soft on an issue to be noncommittal (Presser and Schuman, 1980). A related issue is whether to use an

FIGURE 8.2 LIKERT SCALE

	Strongly Disagree	Disagree	Agree	Strongly Agree
This book is well organized.	1	2	3	4
This book is well written.	1	2	3	4
This book is well illustrated.	1	2	3	4

"undecided" or "no opinion" category. People without a real opinion may feel pressured to express a view if no explicit undecided category is available (Bishop et al., 1980).

There is also no standard rule about how many response categories to use. We noted in Chapter 6 that if too few categories are used, information will be lost. Some respondents will be averse to using the end categories, which also suggests that more categories are better. In scales where the respondent is asked to rate items, as in Figure 8.2, anywhere from between four and seven categories are acceptable. Using fewer than four response categories results in lost information, while using more than seven exaggerates the decision-making abilities of respondents (Miller, 1956).

The consistency of direction in the statements used in Likert scales is another contentious issue. In the example given in Figure 8.2, positive ratings would always be given by circling 4, negative ratings by circling 1. An argument can be made that some statements in the list should be reversed. For example, the second statement might read, "This book is poorly written"—a reversal of direction that would guard against a yea-saying or nay-saying bias. However, do not reverse only a few stray items. Reversed items should be evenly balanced with other items to prevent unintended answers on the part of careless respondents.

Researchers differ on the question of how many categories should be labelled. One strategy is to use a seven-point scale and label only the end points. For example, the extremes might be labelled as 1 ("strongly disapprove") and 7 ("strongly approve"), with no descriptors appearing for numbers 2 through 6. No matter how many descriptive labels are used, design them to help people differentiate between categories (e.g., labels like "excellent" and "good" are better than "very good" and "good").

Finally, the numeric values attached to response categories are also a critical issue. Schwarz et al. (1991) show that using negative number labels in response categories alters the meaning respondents attach to the categories. In their study, people given scales labelled from 0 to 10 generally rated their satisfaction with life as low (with "0" being "unsatisfied" and "10" being "very satisfied"). People given scales labelled −5 to +5 generally rated themselves as more satisfied with their lives. Schwarz et al. concluded that people are less likely to give themselves a negative rating because a rating of "−4" on one scale seemed much worse than a rating of "1" on the other scale, even though the numbers were technically equivalent.

SUMMARY AND QUESTION CHECKLIST

Questioning is an art as well as a science. Good survey research demands a careful crafting of questions to obtain high-quality, systematic evidence. Your choice of words, their order, the form of the question, and the design of the response categories all have important effects on how people respond. In framing questions, the trick is to keep your eye on the overall impression all the questions create while remaining sensitive to the small details of each question. Finally, with every question, it is useful to run through the following checklist:

1. How is this question relevant to the research objective(s)?
2. Do respondents have the knowledge, opinions, or experience necessary to answer the question?
3. Is there a standard question used by other surveyors?
4. Is the vocabulary appropriate and precise?
5. Does the question make grammatical sense?
6. Is the question phrasing sensitive to different cultural or subcultural meanings?
7. Is there a single, clear focus?
8. Does the response format fit the question?
9. How exactly will the answers be analyzed?

EXERCISE 8.1 CONSTRUCTING QUESTIONS

1. For each of the variables in your variable list, construct a question (or set of questions) to measure the variable. If the question is closed, provide an appropriate set of response categories. List the variable each question is intended to measure.
2. Order the questions sequentially to provide what you see as a logical flow.

FURTHER READING

Eichler, M. (1991). *Nonsexist research methods: A practical guide*. New York: Routledge.

Foddy, W. (1993). *Constructing questions for interviews and questionnaires*. New York: Cambridge University Press.

Tanur, J.M. (1992). *Questions about questions: Inquiries into the cognitive bases of surveys*. New York: Russell Sage.

Tourangeau, R., and Smith, T.W. (1996). Asking sensitive questions: The impact of data collection mode, question format and question context. *Public Opinion Quarterly, 60*(2), 275–304.

Chapter 9

ORGANIZING AND FORMATTING THE QUESTIONNAIRE

Having crafted a research focus and proposed questions for all the key variables, we now need to organize the sequence of questions and format them to facilitate responses. The same basic principles used in organizing essays or term papers are applicable to the sequencing of survey questions. There must be an intelligible, logical flow, with smooth transitions between question topics. Furthermore, but more in keeping with a well-structured examination than an essay, there must be clarity in the expectations for answers. Formatting and organizational flow work in tandem to make it easier for respondents to answer the survey questions.

ORGANIZING THE FLOW OF QUESTIONS

Overall, the survey is organized by topic sections. Within the sections, questions should be sequentially organized using a few guiding principles: general to specific, easy to difficult, and less to more threatening. Questions within sections are sometimes organized chronologically from the past to the present or the present to the past, particularly when measuring the history of phenomena. This chapter provides examples of each of these organizing principles.

USING SOCIAL CONVERSATION AS A GUIDE

A survey is a form of social conversation between the surveyor (or interviewer) and the survey respondent (a stranger). In all cultures, conversations with strangers follow well-defined and commonly understood conventions that help people acquire information about one another. Conversations between strangers have a common logic to them, a logic that can be utilized in survey research to provide coherence to the process. Well-organized surveys take advantage of this logic to help guide the respondent systematically through the process of reporting information.

Moreover, it is important not to be careless in constructing your survey. As in a conversation, the respondent will take what you say and how you say it as meaningful to the interaction. The respondent picks up cues from you—cues signalled by question sequencing and formatting, as well as by word choices and order (as we saw in Chapter 8).

The order in which questions involving two main survey topics are asked consists of the following stages:

1. The opening (introduction)
2. Exploring the relationship (characteristics of the respondent)
3. Building up to a main topic (opening questions, including justification)
4. The main topic (detailed questions)
5. The transition (questions to bridge between topics)
6. The buildup to next main topic (opening questions, including justifications)
7. The next main topic (detailed questions)
8. Transition to the end (questions wrapping up final topic)
9. Closing the interview (general questions related to the survey, and closing remarks [thanks and goodbye])

When the survey consists of more than two main topics, extra stages can be added to the process. For each additional topic there is another transition, another buildup, and more detailed questions.

Self- or Interviewer-Administered Surveys

In applying the logic of conversations to the survey process, the specific details of each stage vary depending on whether the survey is self-administered (either on paper or via the Internet) or interviewer-administered. If it is self-administered, the surveyor's part of the interaction has to be predefined and complete; he or she must recognize that there is no opportunity in the self-administered questionnaire to

1. personally motivate the respondent to participate,
2. vary the approach in response to the feedback from the respondent,
3. elaborate a question or statement,
4. probe for further information, or
5. guide the respondent through complex measurements.

With a self-administered questionnaire, the pattern of the conversation must be worked out before it occurs, and all the potential problems must be anticipated in organizing the question flow. An interviewer-administered research instrument, by contrast, allows for each of these possibilities. This flexibility can, however, lead to problems of comparability or non-shared meanings across a sample of respondents. If the interviewer is careless in redefining or restating questions, or in probing for additional information, then different respondents will be responding to slightly, or perhaps even vastly, different questions.

The increasing use of the Internet to deliver surveys allows for greater flexibility in question organization. Unlike a paper-based survey where the questions must be as printed, an Internet-based survey can have the questions tailored depending upon how a survey participant responds. The alteration can involve not only question wording and sequencing, but also the number and focus of the questions asked. For example, if you respond by noting that as a student you are a varsity athlete or a member of student government or you have a part-time job, then tailored questions may be asked about each of these different roles (something that can be done only to a limited extent on a paper-based survey). This type of flexibility is illustrated in several places throughout this chapter.

THE SECTIONS OF THE SURVEY

The Introduction

All surveys begin with an introduction. The introduction may be made by an initial letter or e-mail explaining the survey (see Chapter 10). For self-administered questionnaires, some of this information is partially repeated on the cover page of the survey booklet or on the splash page of a survey website, as indicated in the following example:

> "You have been selected to participate in a research study on _______ . The questionnaire is designed to measure _______ . We believe this research will help us in our understanding of _______ . The research is being sponsored by _______ and conducted by _______ . The survey will take about _____ minutes to complete. Your participation in the survey is voluntary and your answers will be confidential."

For interviewer-administered questionnaires (face-to-face or telephone), the following is more appropriate:

> "Hi! I'm _______ , the interviewer from _______ . We sent you a letter/e-mail a few days ago about our survey on _______ . The survey is about _______ . Your participation is voluntary and your answers will be confidential. It will take about _______ minutes to complete. [If this is a face-to-face interview:] Do you have a place where we could talk without being interrupted?"

In both examples, the introduction includes the purpose of the survey, who is carrying out the survey, a statement about the voluntary nature of the respondents' participation, the confidentiality of their participation, and the time involved.

The Characteristics of the Respondent

In conversation with strangers, people usually start by asking general questions that allow them to categorize or evaluate the other person. Such questions help to define common ground in statuses, attitudes, values, interests, and experiences. In addition to providing a topic for discussion, these questions help us to avoid offending others. For example, at a typical social gathering, the conversation may start with "How do you know the host/hostess?" The answer ("through work," "used to be a neighbour," "we belong to the same gardening group") defines the next question, the answer to which defines the next inquiry, and so on until each participant has built up an image of the person with whom she or he is talking. In general, people are curious about where the other person lives, what education he or she has, and whether that person is married and/or has children.

Surveyors can use the same strategy in their survey instrument. To start the conversation, they can follow this order of questions when trying to determine

- members of the household,
- characteristics of household members,
- characteristics of the dwelling unit, and
- general characteristics of the respondent (age, birthplace, education, income, job).

FIGURE 9.1 EXPLORING A RELATIONSHIP

This is a standard opening used by Statistics Canada in many of its surveys, especially where one of the household members must be chosen as the respondent (i.e., in situations where interviewers phone or visit a household, not knowing the occupants). The demographic questions link well to the questions that are designed to randomly select a household member as a respondent (see Chapter 12). They can also lead to further questions about the household, such as residential history, household expenditures, household recreation, dwelling-unit characteristics, and so forth.

Many surveys (particularly market surveys) leave socio-demographic questions to the end, since respondents may be suspicious of the survey and concerned, at least initially, about disclosing personal information. By telling respondents what the survey is about, the survey introduction creates expectations about what the questions will cover. For several reasons, it is helpful for respondents to get to these central questions quickly. First, it helps keep respondents "on task," setting a professional tone and focus. Second, some respondents will be suspicious or wary, and getting to the point quickly eases apprehensions. Finally, time is a precious commodity and should be used wisely by getting to the core questions as soon as possible. For these reasons, the surveyor may decide to place socio-demographic details at the end or, alternatively, to split socio-demographic questions into two sections. Thus, the questionnaire can begin with household and dwelling details, thereby

FIGURE 9.2 USING SECTION HEADINGS TO PROVIDE TRANSITION

SECTION B **FINANCIAL CONCERNS** This section addresses the factors that are of concern to you in your financial transactions.

9. Please check the category that best applies to you for each of the following statements.

providing the surveyor with an easy introduction for respondents, and conclude with more personal items, including income and birthdate.

Bridging to a Main Topic

Section headings with short descriptions provide visual cues; they signal that a new topic is being introduced. Figure 9.2 shows an example of a section heading that was part of a banking survey. The heading and the short description alert the respondent to the nature of the topic the next questions will address. Some transition questions may also be used to make the conversation flow smoothly from topic to topic (e.g., Now we'd like to focus on banking; could you tell me whether you have a personal bank account?).

Organization of a Main Topic: Easy to Difficult

Opening questions in the first section should be easy for the respondent to answer, salient to the main topic, and non-threatening. At this stage, the surveyor has yet to build trust with the respondent. If respondents are not going to complete the survey, they will make that decision early, so getting to the point quickly is advisable. In the case of a survey on community parks and recreational programs, where the first section asks for information about the use of community recreational facilities, the surveyor could start by asking respondents what they like best and least about living in their community. This section would include questions on parks and open spaces, as well as recreation services and facilities. These could be followed by questions asking respondents to judge the need for additional recreational facilities. The third part of the section could focus on asking respondents to recall their use (or not) of specific facilities. In this flow of questions, the surveyor begins with relatively easy, general questions and progresses to more specific questions requiring more thought on the part of the respondent.

The Transition between Topics

In order to direct the respondent's attention from one topic to another, a bridge is needed. The bridge can take the form of one or more questions. In the recreation survey, to provide a bridge to the second main topic on recommended sources and levels of funding for additional facilities, one question or several questions can be asked about whether the respondent would have used additional facilities had they been available. This eases the respondent from one main topic to the next.

The Buildup to the Next Main Topic

When introducing the next main topic, the surveyor starts with statements justifying why she or he is interested in the topic. For example, in a marketing survey, after asking a number of questions about grocery purchases, the surveyor may introduce a section on socio-demographics by explaining why this information is relevant:

> In order for manufacturers to understand what products consumers want, it's important for them to know more than how much is sold. They need to know who is buying what products and why they are buying them. This survey can help.

The surveyor then shifts to introductory questions on the next main topic by developing respondent interest in the topic.

The Next Main Topic: Organizing Chronologically

A marketing survey on housing demand might include a section designed to determine how the housing needs of the respondents have changed as a result of changes in the family life cycle. The surveyor could start with the respondent's current dwelling, determine when he or she moved into it, and then ask about previous dwelling places and dates, continuing to work backward in time. Using a chronological order—either from the past to the present or from the present to the past—helps the respondent recall the necessary information.

Chronological organization can also be used to determine respondent behaviours that occur through time (e.g., work history; time spent studying; number of friends, relatives, coworkers, or neighbours visited over a given period of time; trips taken; use of different modes of transportation during the course of a day or week; etc.). (See "Section E" on work history in the appendix for an example of time sequence questions.) This process involves getting the respondent to focus on an event such as shopping, having her or him recall the most recent occurrence and provide the necessary details, then working back to the prior occurrence of that event, detailing it, and so on.

This approach may provide more accurate data than asking the respondent to report on her or his shopping behaviour during a typical week, when "typical" may not exist or be readily defined. The surveyor can close the last main topic by asking a few simple questions. Here the concept of typical can be used not to define the pattern, but to determine whether there was a variation in it; for example, a question such as "Would you consider last week typical?" could be used.

Closing

It is important to allow the respondent to state some conclusions about the topic(s) explored in a survey. This practice helps offset any negative feelings the respondent may have about being closely probed and questioned. Some surveyors favour using an open-ended question at the end of the survey, allowing respondents to add detail or provide pertinent comments. It is important that respondents do not feel they are being used. The surveyor should explain to them that their contributions to the survey have been helpful, answer any questions they may have, and thank them for their time.

SUMMARIZING ORGANIZATIONAL ISSUES

The organization of a questionnaire is analogous to the organization of a conversation with someone you have just met. This analogy provides a framework in which we all have some experience. Just as in a conversation with a stranger, the survey goes through the process of opening the conversation, getting to know the person, introducing a main topic, exploring the main topic, making a transition to a new topic, exploring the new topic, and so forth until the conversation is completed. Using this analogous process as a model assists the surveyor in developing and organizing the survey.

Even with interviews, the survey "conversation" stays within prescribed boundaries. However, a self-administered questionnaire is a particularly programmed interaction—a one-sided, controlled conversation that requires the surveyor to anticipate (through pre-testing) all problems. Self-administered questionnaires, whether on paper or on the Web, lack the flexibility of an interview schedule, where a trained interviewer can adapt to potential problems in the organization of the conversation.

FORMATTING THE QUESTIONNAIRE

The instructions, questions, and answers in a questionnaire need to be formatted so that they are relatively simple for the respondent and interviewer to follow. A secondary formatting concern is to make it easy to quickly and accurately code or process the respondent's answers to the questions. Formatting is particularly important in a questionnaire that is self-administered (paper-based or Web-based). If the layout and overall design of the questionnaire are easy to follow and attractive in appearance, then the respondent is more likely to take the time to fill it out.

Formatting is also a concern for interviewer-administered questionnaires, particularly for the first few interviews. The more familiar interviewers become with the questionnaire, the less they rely on the format of the questionnaire to guide them through the interview process. However, if the layout of the questionnaire does not function as an adequate guide, the quality of the initial interviews may suffer, especially if many different interviewers are used.

With computer-assisted interviewing, whether it is face-to-face or via the telephone, questions to respondents and interviewer instructions can be displayed on a monitor one question at a time. Computers allow the interviewer to change the flow of the questionnaire depending on people's responses. The order of questions in specific sections can also be randomized by the computer to prevent response effects that are artificially created by the order in which questions are asked. Internet surveys can also be designed to have the questions sequenced differently for each respondent, either as a function of specific answers or randomly, to avoid order effects.

Formatting Close-Ended Questions

We start formatting at the question level. Close-ended questions consist of three parts: the question itself, the instructions on how to answer the question, and the response categories. Each response category has a value listed and a box, line, or other graphic device for the interviewer or respondent to mark by hand or computer. The easiest form of question to answer and to code is one in which the responses are organized in a column (see Figure 9.3).

A second form of handling responses is to organize them either horizontally (Figure 9.4) or in a horizontal and vertical fashion (see Figure 9.5). These latter methods save space, which may provide a cost-saving benefit for mail-out questionnaires, where weight affects postal costs. However, the overall readability of the questionnaire may suffer, and this must be balanced against any cost savings. With computer-assisted questionnaires, formatting must be sensitive to the characteristics of screen displays.

Various devices should be used to differentiate between the questions and the answers to the questions. For example, questions can be set in one font and answers in another. Questions can also be set in a larger typeface than answers, or they can extend across the page or screen while answers are indented from

FIGURE 9.3 ORGANIZING RESPONSES VERTICALLY IN COLUMNS

In what type of dwelling do you live? *(Please check one)*

House ○
Duplex ○
Townhouse ○
Apartment ○
Other ○

FIGURE 9.4 ORGANIZING RESPONSES HORIZONTALLY

In what type of dwelling do you live? *(Please check one)*

House ○ Duplex ○ Townhouse ○ Apartment ○ Other ○

FIGURE 9.5 ORGANIZING RESPONSES HORIZONTALLY AND VERTICALLY

When there is a disagreement at home, in what language is the disagreement expressed? *(Please check the appropriate category)*

English ○	French ○	German ○	Italian ○	Spanish ○	Other European ○
Chinese ○	Japanese ○	Punjabi ○	Other Asian language ○	Other not listed ○	

both margins. The availability of low-cost personal computers with sophisticated word processing or desktop publishing programs and laser printers has placed the production of attractive print questionnaires within the reach of all survey firms. Larger survey firms have specialists with computer graphics training to graphically design the questionnaire for print or for the Web.

If the researcher has a computer with desktop publishing capability, boxes or circles are easy to create and paste into place. Visual guides (such as rows of dots) leading from the response categories to the boxes to be checked help respondents identify the correct box. These guides should be used when there are a large number of response categories, as in Figure 9.6, or when the category names vary in length.

In the case of paper surveys, the answers can also be pre-coded; that is, the numerical coding values that will be used later in entering the data into a computer can be placed on the questionnaire as superscript or subscript numbers, as shown in Figure 9.7. This allows data to be input directly from the questionnaire into the computer (see Chapter 13). With computer-assisted interviewing or with the use of Web-based surveys, responses are automatically captured in pre-formatted databases.

Linked Questions

Many surveys have questions that are "linked," that is, dependent on the answers to previous questions. Asking respondents which candidate they voted for in the last election presumes that they voted. The respondent should first be asked whether he or she voted. This question acts as a filter or screening question; instructions direct the respondent (or interviewer) to skip to the next appropriate question.

FIGURE 9.6 USING VISUAL GUIDES FROM CATEGORY NAME TO CHECK RESPONSE

Please check all of the interests or hobbies in which you have participated in the last month. *(Please check as many as apply)*

archery	❐	jogging	❐
arts and crafts	❐	lacrosse	❐
baseball	❐	lawn bowling	❐
basketball	❐	lawn darts	❐
badminton	❐	photography	❐
boating	❐	picnic outings	❐

FIGURE 9.7 PLACING NUMERICAL CODES FOR RESPONSE CATEGORIES

Did you vote in the last federal election? *(Please check the appropriate box)*

Yes ❐ 1 No ❐ 0

The formatting strategy is to visually guide the two categories of respondents (the voter and the nonvoter) along different paths either by the visual cues on the paper (see Figure 9.8) or by visual cues on the computer screen. Respondents to whom the follow-up question does not apply skip automatically on the computer or are visually guided to the next appropriate question on the page, while those to whom it does apply are shown or guided to the sub-question(s). Some surveyors have both groups of respondents go to the follow-up question, where respondents to whom the follow-up question does not apply check a box at the top of the response categories that states "does not apply" (see Figure 9.9). In computerized surveys, a set of decision rules can be built into the software to ensure that the proper, linked sequence of questions is followed. In the case of self-administered questions, having the linked questions on one page or screen allows the respondent to visually see all of the linkages.

Repeating Questions

Another type of formatting occurs when questions about a particular topic are repeated many times for different situations, activities, interests, or individuals. This is the question-and-answer matrix. For example, if we needed several items of information about different household members, we could either keep repeating the same series of questions for each household member or we could

FIGURE 9.8 USING VISUAL CUES FOR DIFFERENT PATHS

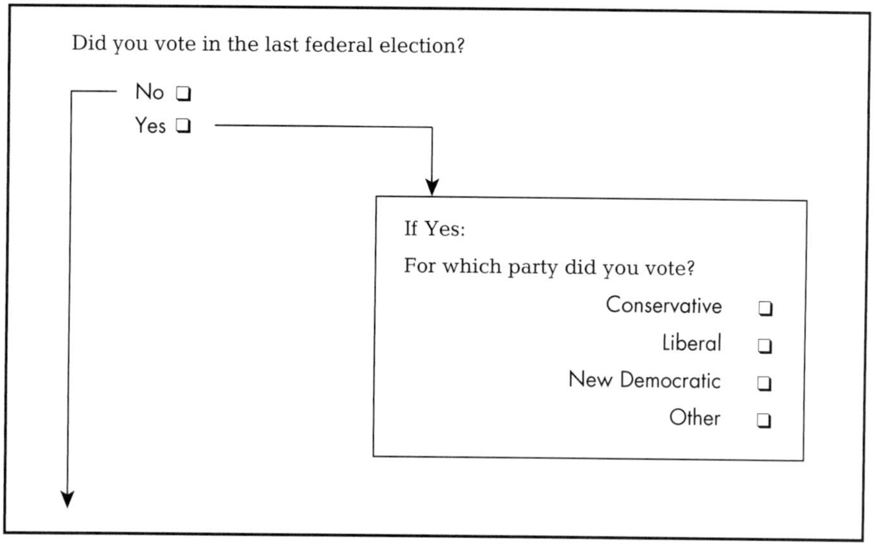

FIGURE 9.9 SELECTIVE QUESTIONING

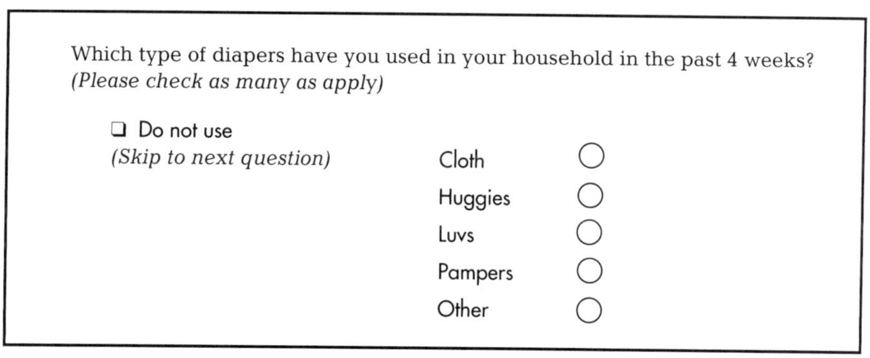

set up a matrix, as shown in Figure 9.10. Using a matrix on self-administered surveys, whether on paper or on a computer screen, reduces the time it takes to fill out the survey, since the respondent can easily see the pattern and is not required to reread instructions for each repeating question.

Although using a matrix simplifies the retrieval and recording of information, only short answers can be recorded because the space is restricted. Footnotes to each category, which specify question meaning and/or answers, are located at the bottom of the matrix. However, matrices such as these may be more appropriate for interviewer-administered questionnaires or for self-administered questionnaires completed by respondents with a relatively high level of education. When the instructions become more complex, respondents

FIGURE 9.10 USING A MATRIX TO HANDLE REPEATING QUESTIONS

First Name of Household Member	Relationship to Respondent*	Birthdate (Month–Year)	Gender	Presently Employed?**
1.______	______	______	______	______
2.______	______	______	______	______
3.______	______	______	______	______
4.______	______	______	______	______
5.______	______	______	______	______
6.______	______	______	______	______
7.______	______	______	______	______

* Write in appropriate category: spouse, child, parent, sibling, other relative, friend, boarder, other (specify).
** Refers to paid work. Include part-time work.

with less education or with less developed language skills may find the questionnaire more difficult to answer.

Repeating Answers

We more frequently have questions in which the answers repeat for successive questions. Opinion and attitude scales are the most common types of questions with repeating answers: that is, the response categories for a number of questions may be yes or no; strongly agree, agree, no opinion, disagree, strongly disagree; good, fair, poor; or satisfied, neutral, dissatisfied. The researcher can visually help the interviewer or respondent, as well as the coder, by creating a matrix with questions as rows and answers as columns, as shown in Figure 9.11.

Fonts, Type Styles, and Type Sizes

Using desktop publishing programs in combination with a laser printer enables the researcher to use various fonts (typefaces), type styles, and type sizes on printed questionnaires. The same choices are available for use on Web pages. Fonts refer to the style of the letters or numbers. They come in two major categories: serif and sans serif. Serif fonts, which are used for the body of the

FIGURE 9.11 USING A MATRIX FOR REPEATING ANSWERS

What additional recreation facilities, if any, do you believe are needed in your community?
(For each facility listed, place a check in the box which best reflects your opinion)

	Yes	No	Don't Know
Gyms	❑	❑	❑
Sports Fields	❑	❑	❑
Picnic Areas	❑	❑	❑
Playgrounds	❑	❑	❑
Tennis Courts	❑	❑	❑
Tot Lots	❑	❑	❑

text, can be distinguished by the fine cross strokes that appear at the top or bottom of each letter. The most common serif font is Times New Roman. Sans serif fonts do not have the fine cross strokes; they are block-faced and have the same thickness at all points. The most common sans serif font is Helvetica. Sans serif fonts are usually used for headings and subheadings, since they attract the reader's attention with their simple form. Sans serif sentences or phrases, however, are more difficult to read; thus, serif fonts can be used for questions, while sans serif fonts can be used for answer categories.

General type styles are regular, bold, italic, and bold italic. Boldface is commonly used for section headings, while italics can be used to differentiate the instructions from the questions.

Type size is defined in point size—72 points to the inch. The larger the type size, the larger the type. Helvetica 18 is a larger type than Helvetica 12. Type size is also dependent on the font. Different fonts vary in length, so that letters set in Helvetica 12 are wider than letters set in Times New Roman 12, which makes Helvetica seem like a larger font. The surveyor needs to experiment in laying out the printed page or the Web page screen of the questionnaire. Once chosen, fonts, styles, and sizes must be used consistently throughout the questionnaire. Consistent formatting for section headings, instructions, questions, and answers helps both respondents and interviewers.

Formatting Open-Ended Questions

Open-ended questions have two parts—the question and the response space. On questionnaires, the formatting of this response space is critical, as it cues the respondent to the surveyor's expectations. Sometimes, as with year of birth or height, the response space can be relatively short: e.g., "In which year were you born?" 19___. With some open-ended questions, an appropriate response unit may also have to be specified: e.g., "How tall are you?" _____ cm or _____ feet, _____ inches. If lengthier responses are appropriate, be sure to provide a sufficient number of ruled lines for respondents to adequately express themselves. Too few lines or too little apparent space on a computer screen will inhibit response if people are frustrated by a lack of space, while too much space might be intimidating to respondents, since this could be interpreted to mean that a short essay is called for.

Too many open-ended questions in a row can be burdensome to respondents, so try to intersperse open-ended and close-ended questions. Occasionally, a short series of open-ended questions may have to be asked sequentially in order to obtain sufficient detail. For example, in asking about a person's occupation, it is necessary to use a series of questions. After establishing that the respondent has a paid, labour force job, it is best to ask at least two open-ended questions in a series: "What is your job title?" and "Please give a brief job description." (leaving a few lines for a response each time). This sequence may then be followed by other open-ended questions focusing on previous jobs or length of time with current employer.

Formatting Sections

Organizing questionnaires into sections that centre on major topics also provides a useful guide to the respondents and interviewers, as we noted earlier. However, there needs to be a sufficient number of questions to make organizing material into a section worthwhile.

The following components are added when formatting the section. As shown in Figure 9.12, a line at the top of the section or Web page is used to signal a new section. The title "Section" is followed by the name of the section

FIGURE 9.12 FORMATTING A SECTION

SECTION C **YOUR PROFESSIONAL ROLE** This section addresses your professional qualifications. The information allows the association to build a profile of our members.

What professional certificates do you hold? *(Please check as many as apply)*

CGA ❑ CA ❑ CMA/RIA ❑

Are you currently a CGA student?

Yes ❑ No ❑

and a short description of the section (including a statement of why the information will be useful). Any new section should start at the top or in the top half of a page or screen, as shown. In this figure, the word "Section" is set in Helvetica Bold 12. The letter C is set in Times New Roman, with a point size of 24. The section heading is set in Helvetica Bold 8, with the body of the instructions set in Candida 8 (questions) and Futura 8 (responses).

Formatting a Page or Screen

It is usually a good idea to design each question separately, and then cut and paste a mock layout for a questionnaire, without filling in all the details. This can be done the old-fashioned way using paper, scissors, and a glue stick, or more easily with a personal computer that has good graphics capability. Work first at the question level, then the page or screen level, the section level, and finally the whole questionnaire.

A typical printed page layout is presented in Figure 9.13. The section heading at the top of the page shows the section title and gives a short description of what the section contains. The first question starts at the left margin and runs across the page. The answer categories are reversed from the usual way of placing "yes" first and "no" second. The reason for reversing them is to guide the people who answer negatively around the subset of questions that apply only to those who answer positively. The arrow leading away from "no" guides respondents to question 30. Those people who answer "yes" are guided to the subset questions beginning with question 28. The subset questions (on company computer facilities) are placed in a box. The box is shaded to visually define a group of questions applicable only to some of the respondents (the shading is light so as not to interfere with the reading of the questions).

Question 30 splits respondents again. Those who have not used a computer are guided to the bottom of the page (to go to the next page); those who have used a computer are guided to question 31. At this question, respondents who do not currently use a computer are directed—as were those who have never used a computer—to proceed to the next page. Those who are currently using a computer are asked questions about location and brand names.

In this page layout, boxes, shading, and arrows are used to guide respondents or interviewers through a series of questions. In setting up such a format system, the surveyor has to be careful that the system does not become unduly complicated or visually confusing. The use of filter or screening questions, skip instructions, and visual cues are crucial, so that people are not asked questions that do not apply to them. The use of different colours can also be helpful, but it should be consistent and tasteful. The same principles apply to formatting a screen image on the Web. However, attention must be paid to the use of different Web browsers that may display information differently for different

FIGURE 9.13 FORMATTING A PAGE

SECTION E **COMPUTERS** The following questions address the issues of computerization of your company/firm as well as the use of computers by the members of your association.

27. Does your company/firm have computer facilities?
(Please check the appropriate answer)
❑ No ❑ Yes

28. What types of computer facilities does your company/firm have?
(Please check as many categories as apply)
❑ Personal computer (stand-alone)
❑ Personal computer network
❑ Terminals tied to mainframe
❑ Mainframe (access limited to specialists)

29. The applications software for your company/firm's computer(s) was:
(Please check as many categories as apply)
❑ Fully custom-designed
❑ Packaged software with major modifications
❑ Packaged software with few or no modifications

30. Have you used a computer? *(Please check the appropriate answer)*
❑ No ❑ Yes

31. Do you currently use a personal computer?
❑ No ❑ Yes

32. Where do you use a personal computer? *(Please check as many as apply)*
❑ at home ❑ at the office ❑ other ______________ *(please specify)*

33. What are the brand names of the computers you use:
(Please write in as many brand names as apply)
at home ______________
at office ______________
other ______________

Please go to Section F.

respondents (see Couper et al., 2004; Fox, Murray, and Warm, 2003; Van Selm and Jankowski, 2006).

FORMATTING AND TYPE OF SURVEY

General formatting can vary somewhat depending on whether the questionnaire is directly completed by respondents or interviewer-administered. The following sections discuss each formatting method in greater detail. One overarching principle is to be consistent. For example, if instructions in one question are in italics, be sure to use italics for all other instructions in the questionnaire.

Mailed Questionnaires

With mailed questionnaires, weight and size are primary considerations in determining postage costs. Weight can be minimized by printing the questionnaire on both sides of the page. The size of the paper used can also affect costs. Using a standard 8.5" by 11" sheet of paper results in the use of large envelopes for mail-out and mail-back purposes, thus increasing postage costs. However, as Figure 9.14 illustrates, turning 8.5" by 14" paper sideways and folding it in half results in a 7" by 8.5" booklet; when the questionnaire is folded twice, it will fit into a standard-size business envelope.

Fitting all the questions on a 7" by 8.5" sheet of paper requires some experimenting with the type size if a laser printer is being used. One important point

FIGURE 9.14 FOLDING A MAILED QUESTIONNAIRE

to keep in mind is that having the type size for certain serif fonts, such as Times New Roman, go below 11 points can compromise readability.

If postage costs are not a consideration, if potential respondents (e.g., seniors) would have trouble reading the smaller typeface, or if many of the questions have subsections and thus require more complex formatting, a larger booklet should be used. In this case, 17" by 11" paper could be folded into an 8.5" by 11" booklet, which requires a 9" by 12" envelope. The costs, number of questions, and number of pages have to be balanced against potential crowding of questions and answers and the overall readability and appearance of the questionnaire.

Self-Administered Questionnaires: Paper or Computer

Self-administered questionnaires distributed via the Internet or hand delivered (e.g., to clinic patients) do not present the same types of concerns as do mailed questionnaires. In the case of paper questionnaires, the decision to use both sides of the page and the size of the questionnaire are determined by the ease with which the respondent can fill out the answers. If the questionnaire is administered in a classroom or office setting, 8.5" by 11" single-sided pages can be used, as they are cheaper to produce, although not as environmentally friendly. However, the larger size of paper, with printing on one side only, may make the questionnaire appear more formidable—that is, larger and thicker—than smaller, double-sided questionnaires.

Although self-administered questionnaires are typically reproduced on paper, computers are now often used as a delivery mode (see Dillman, 2000; Dillman and Bowker, 2001). Some surveys are provided on the Web or sent by e-mail. Some respondents are asked to complete questionnaires that are on personal computers (e.g., when visiting a doctor's office or at a human resources office). While there are technical issues with the use of Web page or monitor displays, a key principle is that the basic formatting decisions used to guide paper-based surveys apply equally to e-mail or Internet surveys (see Groves et al., 2004; Van Selm and Jankowski, 2006).

Interviewer-Administered Questionnaires: Computer

Large survey organizations such as Statistics Canada or major polling firms typically use computer-assisted interviewing. Whether the survey is administered by telephone or in person, the instructions, questions, and answer categories (for closed questions) are sequentially displayed on the computer screen. Here the ease with which interviewers can read from the computer screen is of paramount importance. Using large print in a coloured typeface makes it easier for interviewers to read questions accurately. Typically, these questionnaires are formatted at the question level, then at the section level, and then for the entire questionnaire.

Interviewer-Administered Questionnaires: Paper

Interviewer-administered questionnaires are usually produced on 8.5" by 11" single-sided pages, which can be easily attached to an interviewer's clipboard. It is very important to make it easy for interviewers to read questions and follow instructions.

FORMATTING THE FRONT AND BACK PAGES

Self-Administered Questionnaires

On self-administered questionnaires, the front page or initial computer screen is the first thing a potential respondent will notice. Besides presenting a professional image, this page or screen should include the following information: the organizational identity and logo of the research firm or client organization; the title of the survey; and the introduction to the survey, including its purpose and instructions for completing the questionnaire (see Figure 9.15). The initial

FIGURE 9.15 FRONT PAGE OR INITIAL SCREEN OF A QUESTIONNAIRE

TCU

BC TEACHERS CREDIT UNION

SURVEY

This survey is being conducted by Eikos Social Research to analyze teachers' relations to B.C. financial institutions, including the Teachers Credit Union. The purpose of the survey is to:

❑ help the TCU assess its role in serving teachers' needs,

❑ determine the factors that lead teachers to select particular financial institutions for different services, and

❑ evaluate new programs the TCU could implement to improve the longer-term financial prospects for teachers.

Additional questions are asked about your teaching and socio-demographic background. The purpose of these questions is to allow the researchers to determine the representativeness of the sample, as well as to allow for generalizations to be made about teachers with different attributes.

FIGURE 9.16 BACK PAGE OR LAST SCREEN OF A QUESTIONNAIRE

TCU

BC TEACHERS CREDIT UNION

We thank you for your co-operation in carrying out this research. We are confident that this information will assist the Teachers Credit Union in developing new financial programs for teachers. Please enclose the completed survey in the pre-addressed, postage-paid envelope, and post it as soon as possible.

Survey prepared by Eikos Social Research

EiKOS SOCIAL RESEARCH INC.
SOCIAL AND DEMOGRAPHIC ANALYSIS

screen or front page should be formatted so that the logo and the title of the survey are easily visible (especially for postal surveys that go in envelopes).

The statement thanking respondents for their participation and the instructions for submitting the questionnaire (on the computer) or mailing the questionnaire back are placed at the end. Logos of the organization sponsoring the research and of the research firm can also be included at the end to enhance appearance (see Figure 9.16).

Interviewer-Administered Questionnaires

With interviewer-administered questionnaires, the information on the front page (or on the initial computer screen) is somewhat different. The organization of the front page is designed to provide information (if known) such as the name, address, and telephone number of the respondent; the introduction to be read to the respondent; instructions to the interviewer; and spaces to record information on the final disposition of the questionnaire. *Disposition* of the questionnaire refers to the result achieved, i.e., completed, incomplete, no contact, refusal, or other problem. All of this identification information can then be detached from the questionnaire on completion to maintain respondent anonymity (see Figure 9.17).

FIGURE 9.17 FRONT PAGE/SCREEN OF AN INTERVIEWER-ADMINISTERED QUESTIONNAIRE

Respondent's Name: ______________________
Address: ______________________
Telephone No.: ____________ H ____________ W

Primary Sampling Area [ ]

Interviewer's Name			Number
Record of Contact	Date Called	Time Called	Disposition
First Contact			
Second Contact			
Third Contact			
Fourth Contact			

When you call for an appointment on the telephone

Hi, Mr./Mrs./Ms. ____________, I am ____________ from the Department of Recreation at U.B.C. Did you receive the letter about the survey we are conducting?

If NO: We have been asked by the Department of Parks and Recreation in South Shores to undertake a survey of residents to help them determine the future demand in South Shores for facilities and programs. Your participation is voluntary and would be very helpful.

I would like to book an appointment to talk with you about your and your household's use of recreation programs and facilities. What would be a good time for you?

If YES: I would like to book an appointment to talk with you about your and your household's use of recreation programs and facilities. What would be a good time for you?

Date interview scheduled ____________ Time scheduled ____________

Thank you, Mr./Mrs./Ms. ____________, I will see you on ________ (day of week)

The back page or final computer screen of the interviewer-administered questionnaire contains a statement thanking the respondent for her or his participation, as well as space for the recording of any statements the respondent wishes to add. It should also contain space for the interviewer's remarks regarding the interview (e.g., respondent cooperation, response-quality estimation).

SUMMARIZING FORMATTING ISSUES

The format of the questionnaire must be easy for respondents, interviewers, and coders to follow. In all types of questionnaires, whether self-administered or interviewer-administered, a good layout can increase the respondent's understanding of the questions. A well-defined flow of questions can encourage concentration, and thus accuracy, on the part of the respondent. Furthermore, a self-administered questionnaire that is neat, attractive, and well-organized has a better chance of encouraging respondents to participate in the survey.

PRE-TESTING THE QUESTIONNAIRE

The questionnaire as constructed above should not be considered the final version. It is necessary to see how it works by testing it on potential respondents. This is referred to as pre-testing the questionnaire. The first pre-test can initially involve interviewing accessible respondents, such as fellow students or colleagues at work, to determine if

1. the wording is understood,
2. the most likely answers to close-ended questions are all included,
3. sufficient space is provided for answers to open-ended questions,
4. the flow of the questions follows a proper sequence,
5. the formatting is easy to follow, and
6. the filter and skip instructions work as intended.

The researcher should set up a form on which the interviewers doing the pre-testing can record concerns about the wording of the introduction, instructions, questions, and responses, as well as concerns about formatting. The researcher can then make changes after compiling the results of the pre-test questionnaires.

The same type of pre-testing could next be done on a small random sample from the population to be studied. Language, style, meaning, and sample heterogeneity are typically population-specific, so pre-tests on students or coworkers are not sufficient. The process of linking the questions, determining the order of questions, and formatting the questions would then continue. A recent advance in pre-testing is the use of "cognitive interviewing" (see Willis, 2004). By intensively probing pre-test respondents about their thought processes, survey researchers can learn a great deal about how survey participants understand the questions and the plausible responses. As well, the surveyor must be sure to pre-test all the procedures to be used in a survey, not just the questionnaire or interview schedule; pre-testing the cover letter, for example, can expose misunderstandings or faulty assumptions.

The pre-test responses are then coded to determine if all the questions differentiate answers. If one or more questions result in the same response for all survey participants, it is likely that the researcher has a constant, not a variable. Usually surveys go through five or six versions before pre-testing, and at least two or three versions after pre-testing has started.

SUMMARY

Organization and formatting are critical to creating a successful survey. Whether self-administered or interview-administered, logic, coherence, and appearance make a big difference to the quality of engagement and responses. Furthermore, before the questionnaire is used on the sample of respondents, question wording, meaning, layout, and flow of the questionnaire must be extensively tested. Questionnaires go through a number of versions before they are ready to be used. This is referred to as pre-testing, and it verifies that the questionnaire works as intended. In important ways this pre-testing ensures that you have followed the principles found in Chapters 8 and 9.

EXERCISE 9.1 ORGANIZED QUESTIONNAIRE

1. Order the questions from Exercise 8.1 sequentially to provide a cognitive and conversational flow.
2. Provide the following items: an introduction, section introductions, all necessary instructions for completing the questionnaire/interview, and a closing to the questionnaire.

EXERCISE 9.2 FORMATTED QUESTIONNAIRE

Using some or all of the ideas presented in this chapter, turn your list of questions into a more visually appealing design. Use arrows, shading, boxes, and other graphic cues to make it easier for respondents or interviewers to follow the proper sequence of questions.

EXERCISE 9.3 REVISED QUESTIONNAIRE

Pre-test the questionnaire on three friends or acquaintances who meet the scope conditions for your research. Record the answers, then revise the questionnaire.

FURTHER READING

Dillman, D. (1978). *Mail and telephone surveys: The total design method*. New York: John Wiley.

Dillman, D. (2000). *Mail and Internet surveys: The tailored design method*. New York: John Wiley.

Groves, R.M., Fowler, F., Couper, M., Lepkowski, J., Singer, E., and Tourangeau, R. (2004). *Survey methodology*. Hoboken, NJ: Wiley InterScience.

Willis, G. (2004). *Cognitive interviewing: A tool for improving questionnaire design*. London: Sage.

Chapter 10

WHAT TYPE OF SURVEY?

When responding to a survey, participants can answer either verbally or in writing. Verbal responses come in interviews conducted either face-to-face or over the telephone. Written responses come from self-administered questionnaires distributed through the mail, via the Internet, or by direct distribution (e.g., handed to students in a classroom, given to patients in a clinic).

Recently computers have become integral in the delivery of questionnaires (computers have always been central to the analysis of results). A significant proportion of interviewing is now done over the telephone and computers play a key role: first in randomly dialling numbers, second in displaying questions on a monitor for interviewers to read, and finally in recording responses in databases. When interviews are conducted in person, computers are also frequently used; the questions the interviewer will ask and instructions for the interviewer are displayed on the computer screen. Increasingly, survey respondents complete self-administered questionnaires via computer (e.g., surveys over the Internet) or use the computer to answer parts of a questionnaire (e.g., when interviewers hand respondents a notebook computer to complete a section of a questionnaire).

Table 10.1 summarizes, in a simplified manner, several of the major distinctions between types of surveys. Whether a survey participant reads the questions and responds himself or herself, or does this with the help of an interviewer, is one distinguishing feature of a survey. A second distinction between surveys involves the level of technology included. Postal surveys or face-to-face interviews have relatively low technology levels as compared to Web-based or computer-assisted telephone interviews. The table is "simplified" because it omits a host of complicating factors (e.g., a face-to-face interview in which you

TABLE 10.1 SIMPLIFIED CLASSIFICATION OF SURVEY TYPES

Level of Technology	Type of Administration	
	Self-administered	**Interviewer-administered**
Lower	Mail or postal survey	Face-to-face interview
	Electronic disk sent by mail or an e-mailed survey	Computer-assisted personal interview
Higher	Web-based survey	Computer-assisted telephone interview

are asked to complete a self-administered page of questions—perhaps sensitive questions; or a Web-based survey in which you are asked to listen to an audio clip (the latter akin to an automated interviewer)).

Distinguishing between self-administered questionnaires and survey interviews is critical because the extent of personal contact between the respondent and the research team creates different challenges and opportunities. Interviews allow for more exchange of information with respondents, but even here exchange is more personal in face-to-face interviews than it is in telephone interviews. However, as with many aspects of survey research, there are strengths and weaknesses in this ability to discuss issues with respondents.

One virtue of exchanging information between the respondent and interviewer is that confusions can be clarified. When questions are misunderstood, they can be repeated or restated; as well, if an answer from a respondent seems vague or ambiguous, the interviewer can ask for more details. With a self-administered questionnaire, questions and answers cannot be discussed further (unless respondents are not anonymous, in which case follow-up telephone calls are possible).

The exchange of information that occurs during interviews is not without risks, however. It takes great skill to restate questions properly. By definition, systematic data collection requires that all respondents be asked the same question—restated questions can easily violate this condition. Here is a standard, tongue-in-cheek example of a restated question, asked of a Jesuit priest:

Question: Do you feel it is proper to smoke while praying?

Respondent: Pardon me, could you ask that again?

Restated: Yes, do you feel it is proper to pray while smoking?

By altering the order in which two words appear, the question takes on an entirely different meaning. The likely response of a Jesuit priest to the first question would be "No, smoking interrupts prayer"; however, the likely response to the restated question would be "Yes, all smokers ought to pray"!

As outlined in previous chapters, the choice and order of words, as well as the inflection given to them, can change how a respondent interprets a question. Interviewing allows flexibility in clarifying meaning, but it does so with a risk. Thus, while there are times when self-administered questionnaires work best, on other occasions either telephone or personal interviews may be a more appropriate choice of survey.

MAIL OR POSTAL SURVEYS

Surveys sent through the mail are impersonal and often confused with junk mail. Since people know that completing a survey takes time and effort, you need to convince people to participate in the study. To achieve this goal, there are three basic strategies to follow: create a good impression, make the task easy, and be politely persistent (see also de Rada, 2005; Gendall, 2005).

Creating a Good Impression

First impressions are created by the package in which the questionnaire arrives and by the appearance of the questionnaire itself. "Cosmetics" are very important. To begin with, the envelope in which the questionnaire arrives should have a professional appearance. The potential respondent's address should appear in high-quality type in the middle of the envelope (a mailing label can be used, or the address can be printed directly onto the envelope with a laser printer). The survey organization's return address is essential, and is often professionally printed on the envelope (not a sticky label). Sending the material first-class is advisable. (See Figure 10.1.)

FIGURE 10.1 THE MAIL-OUT ENVELOPE

EiKOS SOCIAL RESEARCH
1661 Duranleau Street
Vancouver, BC V6G 3S3

< Title > < First Name > < Last Name >
< Address >
< Municipality >, <Province > < Postal Code >

The appearance of the mailed questionnaire can influence whether people will respond. Put some care into the front page. Use an appealing, professional layout with crisp summaries of the survey's purpose. Often this is the page that persuades (or fails to persuade) the would-be respondent to answer the questions (see Figure 9.15 for a previous example).

A letter of introduction should be included in the initial mailing. This letter should strike a balance between informing people of the purpose of the survey and encouraging their response. It should be informative and persuasive. Use official letterhead and, if possible, personalize the letter by using the respondent's name (e.g., Dear Ms. Bachman). Figure 10.2 provides an example of a letter of introduction. In designing a cover letter, avoid the appearance of junk mail (i.e., mail that appears to be "form printed" or otherwise mass-produced). The cover letter must also answer a series of questions, listed in Figure 10.3, that potential respondents will almost invariably ask themselves.

Research evidence suggests also that offering an incentive will help persuade people to participate (Goyder, 1987). For example, the Bureau of Broadcast Measurement (BBM) sends people a token amount of money to encourage cooperation. Others try to induce responses by offering free products, lottery tickets, sales coupons or vouchers, and so forth. Giving respondents a reasonable deadline is also advisable. A simple phrase in the letter of introduction, such as "Please return the completed questionnaire by ________ ," encourages people not to procrastinate.

Making the Task Easy

Clear instructions and a simple layout of questions are often the deciding factors in whether people fill out questionnaires. As a first step, respondents frequently skim through questionnaires and then arrive at a judgment about the ease of completion. Physical features influence their decision: Is it easy to read? How long is it? You should also make it easy for respondents to return the questionnaire. Provision of a postage-paid, return-addressed envelope is essential (see Figure 10.4), since only the most highly motivated respondents will supply the postage themselves.

Since first-class postage stamps on unreturned envelopes are wasted, an alternative is to print business-reply postage permits on each return envelope. The cost savings may be large, especially if response rates are low, because you pay the postage only for questionnaires actually returned. This technique requires applying to the post office to obtain the necessary permit and having the appropriate symbols printed on the return envelope. This is, however, easier than licking and sticking hundreds or thousands of stamps!

FIGURE 10.2 COVER LETTER FOR A MAIL SURVEY

EiKOS

EiKOS SOCIAL RESEARCH Inc.
Social and Demographic Analysis

May 6, 2002

< First Name > < Last Name >
< Address >
< Municipality >, BC

The B.C. Teachers Credit Union, as a teachers' organization, serves a significant although specialized population in the province. As such, it wishes to monitor its role and performance in meeting the financial needs not only of its members but of all teachers. Given that the financial situation of teachers in B.C. has remained static for the past several years, finding ways to better serve teachers is even more important.

Eikos Social Research has been commissioned to undertake a survey to evaluate: the degree to which the services presently offered by the B.C. Teachers Credit Union meet the financial needs of teachers; the financial needs of teachers that are not being met by the TCU; the key factors that lead teachers to select a particular financial institution; and new programs that TCU could implement that would improve the longer-term financial prospects for teachers.

You are one of a small number of educators who have been asked to provide their perceptions, data on their use of banking services, and opinions on new programs. You were selected in a random sample of current and retired educators. In order to ensure the sample is representative of educators throughout the province, it is important that each questionnaire be completed and returned. It will take you about ten minutes to answer all of the questions

You can be assured of complete anonymity in providing the data. The questionnaire is not individually identified. Please return your questionnaire to Eikos Social Research in the pre-addressed, postage-paid envelope. Please mail the enclosed card separately.

The results will be used by the TCU in evaluating existing programs and in developing new programs. Highlights of the results will be presented in our monthly news.

We will be happy to answer any questions you might have. Please write or call me at 681-5168 (toll-free 1-800-9-SURVEY).

Thank you for your assistance,

George A. Gray, Ph.D.
Project Director

1661 Duranleau St., Vancouver, BC V6G 3S3 (604) 681-5168

FIGURE 10.3 WHAT TO COVER IN A COVERING LETTER

1. What is this about?
2. Who wants to know?
3. Who will benefit?
4. Why me?
5. How long will it take?
6. Will it be confidential?
7. What do I have to do?

FIGURE 10.4 THE QUESTIONNAIRE RETURN ENVELOPE

Business Reply
Mail

No postage stamp necessary
if mailed in Canada

Postage will be paid by:

EiKOS SOCIAL RESEARCH
1661 Duranleau Street
Vancouver, BC V6G 3S3

Persist, Persist, Persist

A single request to potential respondents is usually not sufficient. Not everyone will return completed questionnaires if you ask only once (they may have lost the envelope, they may think their response is unimportant, etc.). However, response rates can be increased significantly if you follow up your initial request with reminders. Indeed, the research literature demonstrates that the single most powerful determinant of response rates is the frequency of contact with respondents, a finding applicable to all survey types (Couper, 2000; Dillman, 1978; Dillman, 2000; Fowler, 2002).

A scenario for increasing the response rate might work as follows. The initial package is mailed. A week later, a simple reminder letter or card is sent to stress the importance of the project and the significance of each individual response (see Figures 10.5 and 10.6). Three weeks after the initial mailing, another copy of the questionnaire with another letter of encouragement is sent.

FIGURE 10.5 THE REMINDER CARD: FRONT

EiKOS SOCIAL RESEARCH
1661 Duranleau Street
Vancouver, BC V6G 3S3

< Title > < First Name > < Last Name >
< Address >
< Municipality >, < Province > < Postal Code >

FIGURE 10.6 THE REMINDER CARD: BACK

May 16, 2002

Last week a questionnaire exploring your financial needs and transactions was mailed to you. Your name was randomly drawn from the roster of teachers in the province.

If you have already completed and returned it to us, please accept our sincere thanks. If not, please do so today. Because it has been sent to only a small but representative sample, it is extremely important that you be included in the study of the results to accurately represent former and present educators.

If by some chance you did not receive the questionnaire, or it was misplaced, please call EiKOS at 985-5036 or (toll-free 1-800-9-SURVEY).

Sincerely,

George A. Gray, Ph.D.
Project Director

Finally, a week after the second package is sent, a final reminder letter is mailed. Figure 10.7 provides a typical schedule of mailing-out waves.

Repeating the initial request is the key. Research evidence shows that the more times you contact potential respondents, the more likely they are to reply. Following up an initial mailing stresses the importance of the project.

While sending a second copy of the questionnaire reinforces in people's minds the importance of the project, it is costly to mail a second package to

FIGURE 10.7 THE SUCCESSFUL MAIL SURVEY: PERSIST, PERSIST, PERSIST

Initial Mailing	1st Reminder Card	Follow-up Questionnaire	2nd Reminder Card
Week 1	Week 2	Week 4	Week 5

everyone (which you must do if respondents are anonymous). One technique is to include a personalized postcard (see Figures 10.8 and 10.9) in the initial mailing, asking people to mail this postcard separately when returning the completed questionnaire. Respondent questionnaires thereby remain anonymous, while the returned postcards indicate who has and has not cooperated (experience shows that very few people return the postcard without also returning a completed questionnaire). If properly explained to respondents, an alternative is to put unique respondent codes on every return envelope to track returns (with assurances that the survey and the envelope will not be linked by the research team). Using either strategy, follow-up material may then be sent only to those who have yet to reply.

Implementing a Postal Survey

To field a postal survey, the researcher will need to go through the following steps (the parenthesized instructions are optional but strongly recommended):

1. (Send an initial or advance letter advising people to watch for a questionnaire that will arrive soon in their mail.)
2. Send an initial mailing. Included in the envelope will be the following: questionnaire; letter of introduction; stamped, return-addressed envelope; (reply card); and (incentive).
3. Send a follow-up reminder—either a postcard or a letter.
4. Send a second copy of the questionnaire (with a package identical to the first mailing).
5. Send a final reminder postcard or letter.

By drawing a chart showing the number of questionnaires returned each day, the surveyor can determine when to send reminders. After the number received each day has peaked, the surveyor sends out a reminder. This will encourage a new batch of returns, and, after the second peak, the surveyor can send another questionnaire, followed by a final reminder for those who have still not replied.

FIGURE 10.8 THE REPLY CARD: FRONT

Business Reply
Mail

No postage stamp necessary
if mailed in Canada

Postage will be paid by:

EiKOS SOCIAL RESEARCH
1661 Duranleau Street
Vancouver, BC V6G 3S3

FIGURE 10.9 THE REPLY CARD: BACK

I have mailed the questionnaire separately.

< Title > < First Name > < Last Name >
< Address >
< Municipality >, < Province > < Postal Code >

Thank you for your co-operation in completing this study.

The surveyor must also keep track of questionnaires that are undelivered for various reasons (no such address, respondent moved, etc.). To calculate the final rate of response, the surveyor should subtract these returned packages from the total distributed. The response rate, which on a good postal survey should be in the 40 to 60 percent range, is calculated as the number of returned and completed questionnaires divided by the number of questionnaires successfully delivered (factors affecting response rates are discussed later in this chapter).

GROUP-ADMINISTERED OR HAND-DELIVERED QUESTIONNAIRES

Self-administered questionnaires are also appropriate for populations such as students at school, patients in the health-care system, or workers in an organization. The surveyor can either distribute questionnaires to a group or arrange for questionnaires to be given to people. For example, patients can be given questionnaires to complete while waiting for their appointments. Special attention needs to be given to questionnaire returns, especially if the intent is for respondents to take the questionnaire away and complete it later. You need to arrange for completed questionnaires to be dropped off at a specific location or mailed back to the surveyors.

ONLINE, INTERNET, OR WEB SURVEYS

The advent of the Internet, a global network of interconnected computers, has expanded opportunities for survey researchers who can use it to send and receive survey information (see Cha, 2005; Couper, 2000; Couper, Tourangeau, and Kenyon, 2004). Statistics Canada estimates that between 1997 and 2004 the percentage of households with Internet access climbed from 30 to 60 percent (Statistics Canada, 2005). Given the federal government's commitment to expanding Internet access for all Canadians, regardless of their place of residence, we anticipate that this figure will approach 80 percent in the near future. However, access and usage are not the same, and it will undoubtedly take longer for all members of a household, and for all households, to use the full set of communication options available through the Internet.

Although the Internet is limited at this time as a medium for contacting a random sample of all Canadians, Internet usage within some groups is at or very close to 100 percent. Groups for which this is the case include, for example: faculty, staff, and students at all colleges and universities; members of many business and professional organizations; and most students in secondary schools. Where Internet usage approaches 100 percent, using e-mail or the World Wide Web (WWW) as a survey tool is a viable option. Of course, the Internet is an appropriate survey medium when the survey population is Internet users themselves.

The simplest method of using the Internet for survey research is to send respondents a text file containing questions, either embedded in an e-mail message or as an e-mail attachment. Effectively, this is similar to sending respondents a paper questionnaire. However, the Internet is cheaper to use and can be substantially quicker in gathering responses than either mail or telephone interview methods. As with other types of surveys, substantial evidence demonstrates that the greater the number of times a respondent is contacted, the more likely it is that she or he will respond (Dillman and Schaefer, 1998).

In sending questions by e-mail, one method is to list the questions in the e-mail and ask respondents to simply use the "reply" function in their e-mail software. The respondents are instructed to type responses to each question in their reply. While this method is simple, it is also very limited because the questions must be short and uncomplicated. Formatting has to be almost nonexistent, and Likert-type scales cannot be easily used.

Sending a formatted questionnaire as an e-mail attachment is also possible, although this will work well only with respondents who are highly motivated and cooperative. Each respondent must be able to open and use the attachment (which often means having a standard word processing program). An immediate problem is that some users may not have the software necessary to open the attachment, or perhaps can open it but lose the formatting when they do so. One way around this is to send the attachment along with the appropriate software (as a PDF file, for example), but this adds to the effort a respondent must make to complete the survey (and the greater the required effort, the lower the response quality and rate). Furthermore, computer users are becoming increasingly wary about opening attachments coming from unknown sources because of the fear of viruses.

If the attachment can be successfully opened, then the responses are typed in by the respondent. Many people find this easy, but some respondents may get frustrated if they do not have good computer knowledge. Finally, the respondent has to know how to return the completed questionnaire to the survey researcher (typically by saving it as a text file and attaching the file to an e-mail response).

A common alternative to sending the questionnaire to respondents is to direct them to a website where the questionnaire resides. This has become the most common form of Internet survey. Using common formatting software, a survey researcher can construct a questionnaire in which respondents click on buttons or boxes, type responses in pre-formatted boxes, and choose response possibilities from a pre-scripted "pull-down" list (e.g., year of birth, province of residence). Upon completion respondents are typically requested to click a "submit" button to send their answers to a database where information is captured automatically (requiring no reentry or re-typing).

Levels of sophistication of these forms are increasing rapidly. Graphic images as well as audio or video clips are now commonly included. Many online questionnaires also have protocols embedded in them to automatically skip to specific questions depending on how the respondent answers particular questions (i.e., they have automatic skip procedures). As well, online questionnaires can be designed with error-checking controls (i.e., does year of birth and completion of post-secondary education fit?). Embedding these "expert systems" provides self-administered questionnaires with some of the flexibility that is a part of interview-based survey approaches. The software programs that allow this increased sophistication are best thought of as *survey authoring tools* that allow survey researchers to more easily use the interactive programming that is becoming available.

Whether the questionnaire is sent by e-mail or made available on a website, there are several specific pitfalls to avoid. Make sure that questions and response categories are visible on one screen and are not split over two or more screens, so that the respondents do not have to scroll up and down. Do not force respondents to answer every question; in other words, make sure the

INTERNET SURVEY OF CANADIAN MEDICAL STUDENTS

An Internet survey of medical students was undertaken in 2001 (Kwong et al., 2002). The researchers assembled a sampling frame by collecting student e-mail addresses from all 16 medical schools in Canada; foreign students and students from Quebec medical schools were excluded. Foreign students were outside the population of interest (domestic Canadian students), and good institutional records were unavailable for Quebec students. Using data from the Association of Canadian Medical Colleges, the researchers concluded that their frame captured at least 95 percent of all registered Canadian medical students outside of Quebec.

A questionnaire was designed and posted at a specific Internet address. All students in the sampling frame were then sent an e-mail invitation to complete the questionnaire (i.e., technically this was a census, not a sample survey). Non-respondents were sent up to two additional reminders to encourage participation. A PIN number, included with the e-mail messages, was used to ensure that only bona fide medical students completed the questionnaire, and that those who did participate did so only once. Participation was also encouraged by offering draw prizes (e.g., a handheld computer) to be won by a randomly chosen respondent from each school. The response rate was 68.5 percent. As responses were entered by students into the Web version of the questionnaire, information was automatically passed to a database that stored data for all of the 2994 students who replied. (For more information, see Kwong et al., 2002.)

software allows people to skip a question and proceed to the next one. Recall that the basic issues of question wording and formatting discussed in previous chapters are just as applicable on the computer monitor as on the printed page.

Also remember that people using the Web do so with different hardware, different Web browsers, and different user preferences. Just because an online questionnaire looks fine on your monitor does not mean that it will look the same on other monitors. This is as true with images as it is with sounds, animation, and even colours. The amount of time it takes to download digital information (especially complex graphics) directly affects the probability of a respondent abandoning the questionnaire before completion (Couper, Traugott, and Lamias, 2001). Using the latest available software is often a mistake because many potential respondents do not have the most up-to-date software, so what they see or hear may differ from what you see or hear.

Implementing an Internet Survey

To begin, you need to spend significant time on the design of the Internet survey, whether the questions will be delivered via e-mail or posted on a website. This design needs to be tested with different hardware and software to eliminate bugs. Typically there are two ways of implementing an Internet survey: have people who visit a specific website be encouraged to complete a questionnaire or, more frequently, have the survey researcher contact potential respondents about a survey.

The same injunctions we offered about mail surveys apply to Internet surveying. A good impression, ease of use, and persistence are critical to success. For example, addressing a potential survey participant by name (if possible) helps when using the Internet, just as it does when using the mail (see Heerwegh, 2005).

When the survey focuses on people who visit a specific website (e.g., visitors to an environmental group's Web pages or to the site of a company selling merchandise online), you need to motivate these people to complete the questionnaire. First, you have to ensure that potential respondents see the appropriate link to proceed to the questionnaire and are encouraged to click on that link. Second, as with any survey, you need to make the questions relevant and interesting to encourage full completion of the questionnaire. Finally, you usually need to provide some assurance about anonymity or confidentiality, since many people have privacy concerns when using the Internet.

When you are contacting people about an Internet survey, the same principles apply as with a mail survey. Clear explanations about the survey are needed. You then have to follow up with reminders. Finally, you might also consider offering some people an opportunity to talk with an interviewer or complete a postal survey.

FACE-TO-FACE INTERVIEWS

In contrast to the impersonality of self-administered questionnaires, surveys using face-to-face interviews provide a personal connection between the research team and the respondents. This greater contact has two principal benefits. First, respondents are more willing to participate when they are asked to do so personally by an interviewer. Second, respondents are more likely to answer all the questions because interviewers can overcome confusion by restating instructions or questions.

However, this personal touch also brings methodological risks. To be able to systematically compare the answers of respondents, the researcher must

INCENTIVES: WHAT WORKS

Some research (Warriner et al., 1996) indicates that monetary incentives improve response rates in mail-out surveys. But what is the most effective kind of incentive to use?

Warriner et al. (1996) used a mail-out survey on environmental concerns in Ontario to test which type(s) of incentives were most effective in increasing response rates. They experimented with 20 combinations of incentives, combining cash incentives of $2, $5, and $10 with an offer to make a charitable donation on behalf of the respondent (at monetary increments of $2, $5, and $10) and/or the opportunity to be entered in a lottery for $200. They compared the response rates of each specific incentive group to those of a control group that received no cash incentive at all.

They found that offering to make a donation to a charity on behalf of the respondent, regardless of the amount, or using lotteries as an incentive, has no significant effect on the response rate. Offered in tandem with cash incentives, lotteries and donations to charities merely reproduce the effects of the cash incentive alone.

Warriner et al. (1996) found that a $2 prepaid incentive raises response rates by 5.5 percentage points as compared to no incentive at all, and $5 raises response rates by 8.2 percentage points. Bumping the incentive from $5 to $10 made little difference. Curtain, Presser, and Singer (2000) demonstrate similar effects in telephone interviews where respondents are randomly selected. Prepaid incentives enclosed with advance letters increased response rates at least 10 percent (but "promised" incentives had no effect).

make sure that everyone is asked the same question. On a postal questionnaire, this is guaranteed because every respondent receives the same package. In interview studies, however, many interviewers may be used, and thus only a few respondents will receive "the same package" (i.e., the same interviewer). Evidence shows that respondents express different views depending on the characteristics of the interviewer (see Fowler and Mangione, 1990). Since the differences are caused by attributes or practices of the interviewer, the effects are known as *interviewer effects*. It is important to reduce the size of these effects.

Interviewer effects may have a significant impact on responses. As Kemph and Kasser's (1996) study of U.S. undergraduate attitudes toward male homosexuality shows, students reported more negative attitudes about male homosexuality when interviewed by a seemingly heterosexual interviewer than when interviewed by a seemingly homosexual interviewer. In another study, Catania (1996) also shows that the gender of the interviewer affects responses on sexual behaviour. Reese's (1986: 571) study of Mexican- and Anglo-Americans shows that the ethnicity of the interviewer "can exert a significant biasing effect on a variety of cultural questions, even over the phone." How the interviewer conducts himself or herself while administering the survey also affects respondents.

Minimizing Interviewer Effects

Recruitment The surveyor's choice of interviewers will have a bearing on both the rate and quality of responses. As a general rule, it is advisable to choose interviewers who match the survey population (e.g., older interviewers to talk with seniors, members of a particular ethnic group to interview members of the same group, etc.). Matching should not be the key criterion, however. Since interviewing involves communication, people who are good listeners and who can ask the right follow-up questions are essential. Good interviewers must be comfortable following a written script and must be able to interpret and record answers correctly.

Often people do not answer questions directly. Respondents are often vague and indirect, especially when they are not sure what is expected. Interviewers must understand what has been said and ask good supplementary questions to obtain any extra information needed. The surveyor should hire good listeners, not good talkers. Dependability is the other key asset of interviewers. The surveyor needs people who will be cooperative, follow directions, and represent the research team responsibly. Since interviewing does not pay especially well, it is helpful if people are motivated by reasons other than financial ones.

Training Talking with strangers and following a scripted sequence of questions is not an experience most people have had. It is not easy to ask questions exactly as they are written while maintaining good rapport with the respondent. Training and practice help in developing good interviewing skills. A key step in the training is convincing interviewers that following the scripted sequence of questions is essential. The survey team will have devoted considerable time to choosing the proper wording and the best order—an effort that will be wasted unless interviewers follow the pre-set script.

Interviewers must also be instructed in how to accomplish a series of tasks. How do they contact respondents and introduce themselves and the project? How do they decide who should be interviewed? How do they record people's answers? What should they say at the end of the interview? All of these tasks are best taught by example and through role-playing. Interviewers should be instructed on how to restate questions (often repetition alone is sufficient), how to probe for more detail, and how to answer possible questions (e.g., "What do you mean by leisure?").

Some surveys also use visual aids. A rating scale might be printed and shown to respondents in an effort to standardize how they respond to attitude items, or respondents may be asked to comment on diagrams or pictures the interviewer has shown them. Instruction on how and when to use these materials is important.

Supervising Once interviewers begin calling on respondents, a supervisor must monitor their performance to discuss and resolve problems. Completed questionnaires should be reviewed to check that procedures are being followed correctly (e.g., legible responses, enough detail, full completion). Cheating is always a possibility, especially since interviewers are not well paid. Interviewers have been known to create fictitious respondents. The surveyor should re-contact a sub-sample of respondents to check that an interview was in fact done. At that same time, she or he should ask a few key questions to validate the responses. This process not only exposes cheating but also provides, by comparing answers from the two different interviews, an estimate of the amount of error that inevitably occurs on any project.

Implementing a Face-to-Face Survey

Prior to commencing the field portion of the research, the survey team must hire and train interviewers. In the course of the training period (usually four or five days), interviewers are introduced to the project, the questionnaire is reviewed in detail, a few example interviews are conducted, the interviewers role-play as interviewers and respondents (both cooperative and less cooperative), field-trial interviews are carried out, and problems and hitches are reviewed. Only after this introduction does the field portion of the project begin.

When the name of the potential respondent is known, it is best to send this person a letter explaining the study and introducing the interviewer who will call. This saves interviewers from having to make a cold contact and enhances the prospect of completing an interview. In household surveys, however, interviewers often know only the address and not the name of the potential respondent. To produce a representative sample, therefore, some procedure must be used to select a member of the household as the candidate for the interview. Failure to do this results in a skewed sample that overrepresents those people who were both at home and available to answer the survey. We deal with sampling procedures in Chapters 11 and 12.

When personal interviews contain open-ended questions, some method of recording answers needs to be considered. Tape recording the interview has several advantages: verbatim detail is captured, the interviewers can give their full attention to the respondents, and all members of the research team can listen to the answers. The drawbacks include the high cost of transcriptions and, most significantly, the invasion of privacy. Most respondents who consent to taping quickly forget the presence of the tape machine once the interview is under way. An alternative is to have interviewers take notes, but this can be distracting to respondents and sometimes results in rushed and illegible notes.

There are a number of administrative tasks to perform as well. The surveyor should provide interviewers with personal identification material, which can be shown to potential respondents in order to verify the legitimacy of the survey and the interviewer. Moreover, the surveyor should devise a payment schedule (per interview or hourly) and a method of compensation for expenses (e.g., travel). Finally, it is imperative to review with the interviewers all ethical issues, including their safety.

TELEPHONE INTERVIEWS

A major innovation in survey research has been the growth of telephone interviewing (see Genovese, 2004). Two things have triggered this increase. First, almost all households have telephones, so a surveyor's use of the telephone does not bias respondent selection as it would have a few decades ago. Second, and most significantly, telephones allow computers to play a much more central role in survey research.

Prior to the advent of the microcomputer, the only part of survey research influenced by computers was the data analysis phase. Today, computer-assisted telephone interviewing (CATI) is standard in the survey industry. The computer plays the following roles in the CATI system:

Dialling The computer places the call, choosing the number either from a predetermined list or by randomly dialling numbers. The computer can also track the number of times a number has been tried and the times of day when calls have been placed.

Script The computer screen shows the interviewer what to read, from the opening script through to the specific questions and the closing thank you. The question order and sequence can be altered by the computer either based on the answers respondents give or by design of the surveyor. The latter case usually occurs with attitude questions when the order of the questions is randomized for each interview (to cancel out any effects that may result from following a certain pre-set order).

Data record By typing answers verbatim or by typing appropriate numbers, the interviewer can record the answers in computer memory as soon as the respondent speaks. Certain typing errors can be immediately caught (e.g., out of range numbers), and inconsistent answers from respondents can be double-checked, since the computer can prompt the interviewer to ask for more information.

Especially important in the CATI system is the concept of random-digit dialling (RDD), which gives survey researchers a relatively simple and comparatively cheap method of securing a randomly selected sample of people. By

FIGURE 10.10 ESTABLISHING RAPPORT ON THE TELEPHONE

instructing the computer to call only numbers in "working banks" (i.e., by pre-selecting area codes and three-digit exchanges used by telephone companies), the last digits of telephone numbers can be randomly generated by computer (see Chapters 11 and 12 for a more detailed discussion).

The political polls we see or hear reported in the media use the RDD method. From a few centralized telephone rooms, people across the country can be interviewed. The disadvantage is that the survey call may be confused with nuisance telephone calls. The growing popularity of telemarketing has meant that many households are being called frequently, and complaints about the invasion of privacy are rising. Success in telephone surveys relies on the ability of interviewers to gain cooperation from respondents. The first few seconds of the call are essential. Respondents must quickly recognize that it is a survey research firm calling, that nothing is being sold, and that their cooperation (for only a short time) would be appreciated. Cellular phones also complicate, but do not undermine, this process (see Keeter, 2006). While RDD methods will catch cell phone numbers, cell phone users are mobile and these calls can catch people while they are driving, shopping, and so forth. In these cases respondent cooperation is difficult to secure, although arranging for a time to call back will sometimes work.

Very similar to CATI is the CAPI system, or computer-assisted personal interview, which is also increasingly popular. With CAPI, interviews are conducted face-to-face, but instead of the traditional paper-and-pencil method, interviewers use laptop computers. CAPI shares many of the benefits of CATI, and as Couper and Burt (1994) show, CAPI is generally well received by interviewers as an instrument with which to work and is often preferred to pencil-and-paper interviewing.

WHICH METHOD SHOULD BE USED?

Group Administration

Passing out questionnaires to groups of people (say students or employees) is usually a good choice because cooperation is often very high. The problem with group administration, however, is that very few survey populations meet in groups. Even when they do, obtaining permission (e.g., from a school board or hospital) to survey the group is often problematic.

Another reason for avoiding this approach might be the nature of the questions. For instance, surveys of workers in their place of employment may be a poor choice if your questions make respondents uncomfortable, possibly raising ethical issues (e.g., if you ask about the boss or coworkers). You need to be sensitive to the setting in which questions are asked.

Choosing the Type of Survey

The choice of the type of survey to use is usually dictated by sample limitations, response rates, cost, subject matter, speed, timing, or a combination of these factors.

Sample limitations The amount and kind of knowledge you have about the study population influences how you can collect data. For postal surveys to work well, you must send the questionnaire to a specific person at a specific address. Telephone interviews that do not rely on RDD obviously require at least a telephone number, and preferably a personal name, or at least a title or position (e.g., club president, personnel director).

RDD requires neither name nor address, although care must be taken to ensure that all working telephone numbers are eligible for selection (only about 2 percent of Canadian households have no telephones). This method is viable for surveys of the general public, but not for surveys aimed at specific people or people in specific positions (e.g., hospital administrators). You also cannot distinguish between business and residential telephone numbers using the RDD method.

Interviewers can be sent to addresses, and the names of the occupants need not be known. Lists of addresses can be generated from city directories, and then interviewers can visit each dwelling and select a household member to interview. New housing developments may be missed on such lists unless special steps are taken to ensure their inclusion. Interviewers may encounter difficulties with apartments and condominiums where security systems impede access, or they may be reluctant to work in certain neighbourhoods.

Internet surveys are particularly susceptible to sampling problems. However, for specific groups (e.g., university or college students, employees of a firm), Internet surveys are becoming more popular and effective. A key issue is access to a complete listing of e-mail addresses for the population of interest. More and more organizations have this information on file, so in the future sample limitations will be less problematic for Internet surveys. Issues of privacy are an issue here though, since many Internet users continue to be wary of "cookies," computer viruses, spyware, and the like ("cookies" are information strings entered into a computer's memory via Web browsers). For general population surveys, however, the Internet is not effective because large segments of the population do not have regular Internet access (see section "Online, Internet, or Web Surveys," above).

Response rates If only a small percentage of the people asked to participate in a study do so, the results are subject to bias. Since those who agree to participate *may* be very different from those who refuse, obtaining high response rates is crucial. For respondent groups whose interest in a survey research topic is high (e.g., surveying high-performance athletes about nutrition and training), response rates will be high no matter which method is used.

Postal surveys typically elicit less cooperation (i.e., lower response rates) than do interview methods in general population surveys, although with good design and persistent effort, cooperation rates can be high in both. Questionnaires for the general public addressed to "householder" or surveys of businesses sent to "owner/manager" result in low cooperation. As well, questionnaires are not appropriate for special populations who may have trouble reading or writing (e.g., seniors, recent immigrants). In both instances, an interview study would achieve greater cooperation and more reliable results.

Telephone surveys—especially RDD surveys—typically have lower response rates than do personal-interview studies. However, telephone and personal-interview studies have similar rates of respondent participation if a letter explaining the study can be sent to people prior to the placement of a telephone call. Finally, telephone interviews may have lower rates of measurement error because the audio-only medium eliminates many of the nonverbal cues associated with personal interviews (where interviewer appearance and gesture have influences). Yet telephone survey response rates have been dropping in recent decades, a trend that caller ID has influenced (Curtin, Presser, and Singer, 2005).

Internet surveys are relatively new, so response rate research is still in its infancy. Early indications are that in the right circumstances—that is, among groups where e-mail access is high—response rates are in line with those for other survey types. For specific population groups, such as members of organizations, there is little difference between the postal and Internet methods, for example, if appropriate strategies to enhance response rates are used (e.g., Dillman and Schaefer, 1998). However, if the survey is of the general population, Internet response rates are low, largely because so many people lack Internet connections. Survey response rates of visitors to websites, where people visiting a certain page are requested to complete questionnaires, are low, and typically such "volunteer" samples are not representative of all visitors to specific websites (see Chapters 11 and 12 for further discussion).

Cost The postal survey is often thought to be a cheap method, but if response rates are to reach acceptable levels, the costs will still be substantial. Professional design costs are high, and multiple mailings can quickly increase costs. Properly executed postal surveys are only fractionally cheaper than telephone surveys.

The price of telephone surveys is driven primarily by wages for interviewers and long-distance charges, although the initial capital costs of establishing a telephone-interviewing centre are high. CATI systems require microcomputers connected to telephone lines and interviewer headsets, as well as the associated computer software.

Personal interviews are more expensive because of the costs of transportation and salaries for the interviewers. Indeed, if respondents are widely dispersed, personal interviewing can be prohibitively expensive.

Web surveys are relatively cheap, although (like postal surveys) when they are well done, they are more expensive than most people think. The questionnaire must be professionally laid out, the time required to load pages must be short, the flow through the questionnaire must be seamless, and the visual display must be carefully designed. Web and graphic designers need to be hired, and the issue of capturing the information in databases must be addressed. However, there are now several firms offering Web-based survey platforms for a relatively low rate (see the website listed at the end of the chapter).

Subject matter Answers may depend upon method. For many people, it may be harder to admit to illegal activity (e.g., shoplifting) in an interview than on a postal survey. Conversely, it can be argued that it is more difficult for some people to lie to an interviewer than to lie on a questionnaire (either on the Web or on paper). This implies that each method may have different effects for different people. Indeed, the survey research literature suggests that the overall pattern of responses, even on sensitive topics, tends to be similar whatever the method of data collection (Singer and Presser, 1989: 187–244).

If the subject matter requires many questions—and especially detailed questions—a personal interview is usually the best choice. The very presence of an interviewer encourages people to continue answering questions, and a trained interviewer can explore complex topics in detail. Many people grow impatient with lengthy telephone interviews (i.e., over 15 minutes), and long questionnaires reduce the likelihood of people even starting them. Telephone interviews often have a fast pace (due to time pressures and possible refusals or early terminations) and responses to open-ended questions are often truncated.

Speed Some research requires fast results. It would make little sense to field a political poll through the mail since the election would likely be over before all the questionnaires were returned. Sending out interviewers is also time-consuming unless you have an extremely large team. Telephone surveys produce quicker results, and Internet surveys are far more time-efficient than are postal surveys.

Timing Completing questionnaires is easier for people at certain times of the year. Mailings to the general public around holiday periods are ill-advised. For various organizations, it is more inconvenient to receive a questionnaire at the end of the month than to receive it earlier in the month. Surveying during July, August, and December is especially hard, no matter which method you choose.

MIXING METHODS

Methods can also be mixed. For example, you might combine a face-to-face interview about family life with a private questionnaire containing sensitive questions about marital relations. To improve response rates from self-administered questionnaires, you might hire people to drop the questionnaires off or to visit people who have not returned their surveys. You might also ask people on a postal survey to supply a telephone number so that questions of clarification can be asked in a telephone interview.

SUMMARY

The type of survey chosen is dictated by considerations of sample limitations, response rates, cost, subject matter, the necessary speed of data collection, and timing. When the essential information requires only a few simple questions, a self-administered questionnaire makes good sense. As the complexity of the information required increases, interviews become more attractive, since interviewers can overcome confusion and motivate respondents to participate. Internet surveys are becoming more popular, but access to representative samples remains a problem. Whatever method is chosen, it is crucial to remember that the survey package is an integrated whole, and the total design influences both the quality and quantity of response.

EXERCISE 10.1 ORGANIZE ADMINISTRATION OF SURVEY

Decide which method of data collection to use for your survey. Provide a justification for your choice, and outline the basic steps you would use to implement your project. If you choose a mail survey or Internet survey, provide a flow chart for the timing of different events. Provide examples of material for all relevant stages (e.g., letters or e-mails, envelopes, reminder messages, etc.). If you choose an interview method, describe how respondents will be contacted. Give a general outline of what interviewers will say to respondents, and provide specific suggestions for how interviewers are to deal with specific questions and objections (e.g., "Why me?" "I'm too busy right now—can my friend do it?").

FURTHER READING

Baker, R. (1992). New technology in survey research: Computer-assisted personal interview. *Social Science Computer Review, 10*(2), 145–57.

Couper, M.P. (2000). Web surveys: a review of issues and approaches. *Public Opinion Quarterly, 64*(4), 464–94.

Dillman, D. (2000). *Mail and Internet surveys: The tailored design method.* New York: John Wiley.

Groves, R. (1990). Theories and methods of telephone surveys. *Annual Review of Sociology, 16*, 221–40.

Groves, R. et al. (1988). *Telephone survey methodology.* New York: John Wiley.

Maynard, D., Houtkoop, H., Schaefrer, N., and van der Zouwen, J. (2002). *Standardized and tacit knowledge: Interaction and practice in the survey interview.* New York: John Wiley.

On the Internet

For a sample Web-based survey tool, see Survey Monkey: http://www.surveymonkey.com/

Chapter 11

SAMPLING FROM POPULATIONS

The next steps in survey design involve decisions about sampling. Constraints of time and cost make it rare that everyone (or everything) in a population can be included in the study. This creates a challenge for surveyors. Although the entire population cannot usually be surveyed, it is this larger population about which conclusions need to be made. For example, in doing research on urban waste disposal, it is practically impossible to interview all city dwellers, even though the intention may be to use the survey results to design policies relevant to all members of that group.

Recall that while all surveys have individuals as respondents, surveys can focus on people or on the groups/organizations to which people belong. So, surveys can include studies of, for example, households, businesses, daycare centres, hospital units, or municipalities. For convenience, we refer to the individual elements of a population as units; thus, if all city daycare centres constitute the population, each separate daycare facility is a unit of the population. Exactly which units to include in the sample involves many decisions. A series of questions can help to guide us through the complex process of sampling (see Figure 11.1).

Practical limitations mean that information from a subset of the population—i.e., the sample—must be treated as representative of the entire population. In other words, information about a population must be inferred from information based on a sample. Therefore, how adequately the sample (e.g., of people, or of daycare centres) represents the target population is critical. This makes knowledge of sampling techniques essential.

FIGURE 11.1 SAMPLING DEFINITIONS

■ Who *is* the target group for the study?	→	This is called the **study population**.
■ Who *in* the target group should be surveyed?	→	This is called the **sample**.
■ How *many* people should be surveyed?	→	This is called the **sample size**.
■ How should the people to be surveyed be *selected*?	→	This is called the **sampling method**.

What kinds of samples are representative? When medical professionals take a blood sample, they poke us only once (hopefully!). They do not need blood samples from many different parts of the body. The reason for this is that one drop of our blood is just like any other: any two blood drops are alike or *homogeneous*. Therefore, health-care workers can confidently draw general conclusions from a single sample of blood.

In survey research, this is never the case because one person cannot represent the views of everyone. Even a few people cannot represent the views of a large group because human populations are not homogeneous (at least not in the dimensions or issues relevant to social scientists). For this reason, survey researchers must ensure that their conclusions rest on a representative subset of the population. The accuracy of a study is only as good as the representativeness of the sample on which the conclusions are based. But how are decisions made about which units in the population should be in the sample? A systematic rule is needed for deciding which units to include and which to exclude—the laws of probability provide this rule.

PROBABILITY SAMPLING

We could do a survey of city residents by asking for volunteers. The result, however, would likely be a very unrepresentative sample. It would contain mainly people who were outspoken and/or who had nothing better to do. People who were shy or busy would not be included, and therefore the sample would not be representative of the general population. What is needed instead is a way of ensuring that everyone has a chance of being included in our sample. Probability sampling does this. The probabilities need not be equal for every unit (as discussed in Chapter 12), but every element or unit in the population must have some likelihood, or a non-zero probability, of being included in the sample.

FIGURE 11.2 PROCESS OF SAMPLING

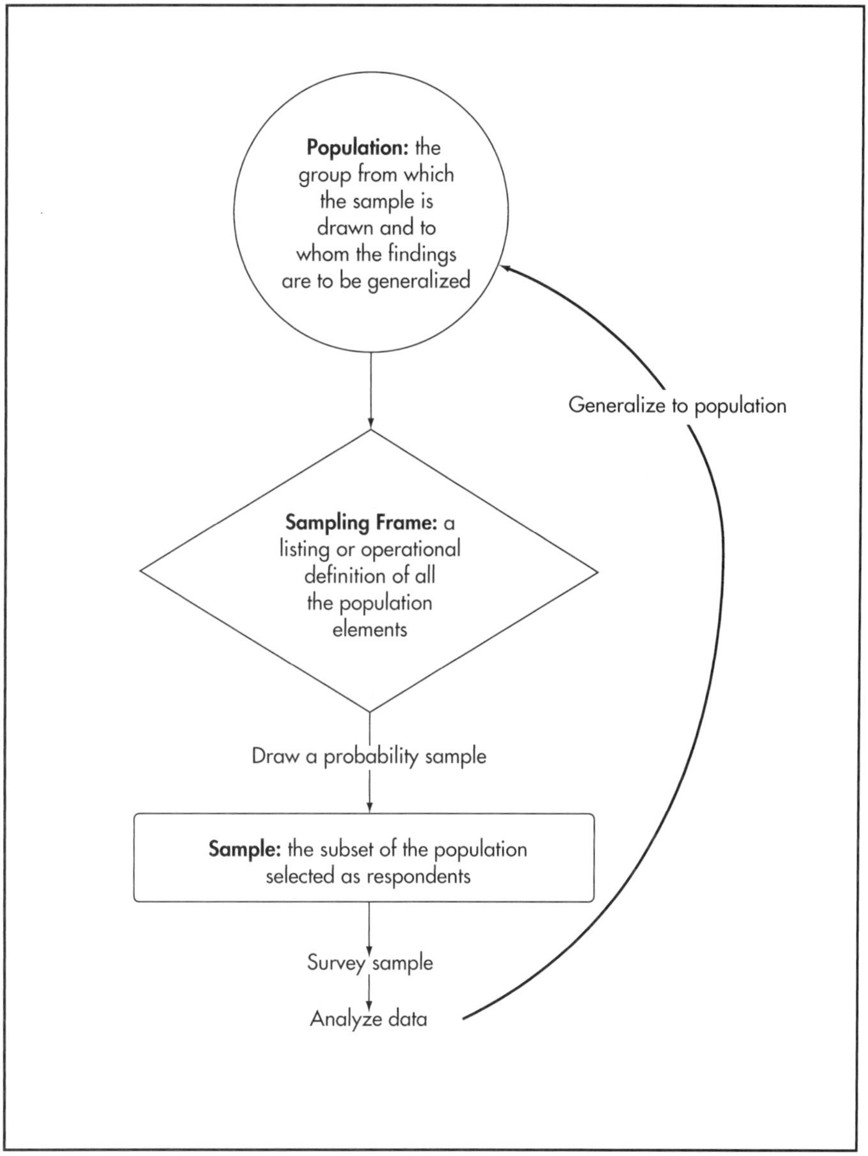

To understand probability sampling, it is important to distinguish between three groups: the population, the sampling frame, and the sample (see Figure 11.2). Think of a research topic involving a survey of all students currently enrolled at your college or university. The target group (the *population*) would be all students, while a listing of students (perhaps the one kept by the college or university registrar) would be the *sampling frame*. The distinction between a population and a list of the population (sampling frame) is an important one. The population of students and a list of students may not be the same (e.g., dropouts). Finally, the *sample* would be a randomly chosen subset of all the *listed* students.

SAMPLING FRAMES

The concept of a sampling frame refers to either a listing or a procedure that includes all the elements or units in a survey population. For example, a *listing* of all the daycare centres in a particular community could serve as a sampling frame for a study of community daycare centres. So long as "daycare" was specifically defined (excludes baby sitters, requires a business licence, etc.), then a sampling-frame list could be made.

In practice, sampling frames rarely provide a perfect match to a population. Given this difference, a key sampling task is to establish a sampling frame that captures the population as completely and accurately as possible. Consider the earlier example of sampling students. In theory, it is easy to visualize a list of all students. However, in practice, a completely accurate list of students is difficult to obtain: students drop out of school, or their names and addresses change or are recorded incorrectly by the registrar. Given that the sampling frame is unlikely to be perfectly accurate, the goal is to establish the most representative sampling frame possible.

While the sampling frame has been traditionally defined as a listing of population elements, recent advances in random-digit dialling (RDD) mean that no full listing of all sample units exists. RDD relies instead on a procedure that, when followed properly, includes all units or elements of the target population (see Chapter 12 for a fuller account of RDD).

SAMPLE SELECTION AND SAMPLE BIAS

The sampling frame provides a means for selecting (using the rules of probability) a representative sample of units. Random selection is the key to the process of choosing a sample from the sampling frame. To return to our school example, we could write the name of each student listed in the sampling frame on a separate slip of paper and then draw name slips at random, thus generating a random sample of students. Under the rules of chance, our sample is more likely to represent all students than if we asked for volunteers, chose our friends, or selected only those people we met in the library. However, our sample will not be perfectly representative. As we have already seen, error will be introduced by our reliance on an imperfect list of students. Sampling frames are rarely perfect listings of populations. The difference between the population

FIGURE 11.3 SAMPLING FROM A HAT

and the sampling frame is called *sampling bias*. Some also refer to this as *coverage error*, in the sense that a sampling frame that is fully accurate should "cover" the entire population.

Reducing sampling bias requires careful attention to the integrity of a list of population units or to the effectiveness of RDD procedures used in targeting a population. In surveys of any general population (e.g., residents of a community or province), lists of people are usually not reliable because people often change residences. Telephone directories, voting lists, property tax rolls, or city directories become outdated very quickly, but they are useful starting points in constructing sampling frames of household addresses. It should be noted that address listings taken from such sources typically exclude recent housing and apartment developments, as well as illegal suites (i.e., undeclared dwelling units not permitted under existing zoning). If these deficiencies can be rectified by sending fieldworkers to enumerate dwellings in selected areas, a more accurate sampling frame of households can be created.

Compared to sampling the general population, it is usually easier to sample members of organizations or organizations themselves, since they tend to keep fairly accurate lists that can be used as sampling frames (see box below for exceptions). For instance, lists of patients who have recently visited a particular clinic, or even lists of clinics, are often very accurate and up-to-date.

SAMPLING BIAS

Sample frames derived from institutional lists are only as accurate as the institutions need them to be. Lists of students or employees are relatively accurate and up-to-date, since these records are essential for the institution. Lists of prison inmates also demand accuracy.

Lists of volunteer agency members or business customers are often not as accurate because there is less need to keep records current. Members and customers move, and these lists become dated quickly unless someone is responsible for the upkeep.

One of the difficulties with using Internet surveys is the extent of sampling bias or coverage error. In Canada, only about 60 percent of households currently have Internet access. Using the Internet for general community surveys would not cover a sizeable portion of the population. However, e-mail lists of students in a university or college might be reasonably complete and accurate. Company employees, members of associations (e.g., professional associations, political parties, voluntary groups) and some organizations (e.g., business groups, technical groups) routinely use the Internet for communication. Sampling from those lists that are accessible would provide adequate coverage, so that sampling bias would be minimal. In these cases, a Web survey might be appropriate (see Dillman and Bowker, 2001, for a good discussion of sampling issues with online questionnaires).

RANDOM SELECTION AND SAMPLING ERROR

There is another reason, beyond problems with sampling frames, that our sample is unlikely to be perfectly representative of the population. Chance determines which units are chosen from the sampling frame. Random selection gives us the best method of obtaining a representative sample, but it does not guarantee a perfectly representative sample. Just by chance, there will be some *sampling error*.

Sampling error can be illustrated by flipping a coin. In a sample of 100 flips, it would not be unusual to obtain a result of 55 heads and 45 tails, even though we would expect a 50–50 distribution of heads and tails in the long run (i.e., if we kept flipping the coin indefinitely). The 55–45 distribution from our sample of 100 coin flips is the result of random fluctuation. In a similar way, chance factors typically mean that samples do not perfectly represent populations. Even if the distribution of students at your school is 50 percent female and 50 percent male, we should not assume that a randomly selected sample will produce an equal number of women and men. However, if we cannot include all population units, a probability sample is the best method of ensuring the most representative sample.

The principle of random sampling is analogous to the idea of shuffling playing cards before dealing. All players in the card game have an equal chance of obtaining a particular card, so long as the deck has been well shuffled. The fairness of the card game, like the fairness of any sample, rests on the laws of probability.

Consider the random sampling that commonly occurs when polls are taken. Often when the results of a political poll are announced, a certain margin of error is also reported: "The results are accurate within +/− 3 percent (*margin of error*), 19 times out of 20 (*level of confidence*)." The message is that the laws of probability reduce but do not eliminate sampling error in survey research.

With any probability sample, we can only *estimate* values in the population. In a survey of 1000 voters, we might find that 30 percent of the sample say they

CHEATING ON THE LAWS OF PROBABILITY

All professional fields use specialized jargon. When reading the results of survey research, don't be fooled when someone uses this jargon inappropriately. Here is an example of statistical bafflegab. A Canadian business consultant asked people how satisfied they were with Internet services. He did this by recruiting 500 "study participants ... from online Canadian discussion forums (i.e., moderated bulletin boards, usenet groups, and opt-in mail lists)." He then reports that "a survey of 500 respondents from a total Canadian population of 31,156,393 ... has a statistical accuracy of +/−4.3%, 19 times out of 20" (McDonald, 2002). First, the 500 respondents do not constitute a random sample and hence the laws of probability on which the margin of error and the confidence level are based do not apply. Second, even if it had been a random sample, it is not a sample representative of the Canadian population because not all Canadians have Internet access. Third, it is extremely unlikely that only Canadians were among the 500 respondents if respondents were recruited from the Internet, where geographical boundaries have little bearing. Beware those who use specialized statistical jargon to persuade readers of outlandish claims.

will vote for Candidate A. The margin of error for this sample estimate is expressed in the mathematical concept of "+/− 3 percent." From a random sample of 1000 voters, we can confidently predict that Candidate A will receive between 27 percent and 33 percent of the vote. However, even though we can be confident, our level of confidence is not perfect, and this is expressed in the idea of being right 19 out of 20 times (or 95 percent of the time). This is often called the *confidence interval*. Exactly how confident we can be about any sample estimate depends upon both the size of the sample and the heterogeneity of the population.

SAMPLE SIZE

The required sample size depends principally on two things: (1) the *heterogeneity* or variability of the population in question, and (2) the degree of accuracy required in conclusions.

Heterogeneity and Sample Size

If the population is homogeneous, like the blood sample discussed above, we can be confident that only one sample observation is sufficient. But if our population is very heterogeneous, a larger sample is necessary if we are to place any confidence in our findings.

FIGURE 11.4 HETEROGENEITY OF A POPULATION

To say with confidence how many toes the average individual has, you would have to examine only a few pairs of feet before being confident that almost everyone has 10 toes. To confidently report the proportion of individuals with brown eyes, however, you would have to observe more than a few pairs of eyes. The difference comes in the fact that almost everyone has 10 toes (people are almost perfectly homogeneous with respect to the number of toes), whereas the colour of people's eyes is much more heterogeneous. Just how many pairs of feet and eyes you would need to sample in order to draw a confident conclusion is a question of sample size. In contrast to conclusions drawn about a homogeneous population, with a more heterogeneous population, a larger sample is required to achieve a similar level of confidence.

Accuracy and Sample Size

The larger a random sample, the more closely will estimates based on the sample approximate the actual values in the population. If you estimate the average age of the students at your college or university after asking a random sample of 50 people, you will be less accurate than if you base your estimate on answers from 200 randomly selected people. In fact, the rules of probability tell us that we reduce our margin of error by one-half if we quadruple our sample size. Therefore, increasing our sample size to 800 students would reduce the margin of error in our estimate by one-half again.

If we were conducting an exploratory, descriptive study, a smaller sample might be tolerated if the precision of our estimates was not crucial (e.g., not to be used as the basis for policy decisions). However, if we wanted to know the likely effects of some crucial piece of legislation, a larger sample would be necessary because it would provide more accurate results. Too often one hears or reads about survey results, possibly from online surveys, that proclaim some finding based on "thousands" or even "tens of thousands" of respondents. Such large numbers of respondents are meaningless in terms of accuracy or scientific validity unless they are representative of the larger population to whom the results are deemed to be applicable. Large unrepresentative samples are of no utility in good survey research.

In the polling results noted above, two measures of accuracy were discussed: (1) the result was taken to be within +/− 3 percent of the true value, and (2) this result was considered to be accurate 19 times out of 20. We will consider each of these claims in turn.

Margin of Error

The measure of +/− 3 percent refers to the margin of error, the size of which is directly related to the size of the sample. For example, consider a survey where people are asked about paying a fee for the collection of excess residential

FIGURE 11.5 MARGIN OF ERROR BY SAMPLE SIZE

If 70% of a sample "agrees," then with	the population value might be	
a sample size of	as low as	as high as
10	34%	94%
50	55%	83%
100	59%	78%
250	64%	76%
1000	67%	73%

garbage (say $5 for every can of garbage beyond two can loads). A survey firm reports that 70 percent of people in the city think this is a good idea. The margin of error in that estimate is critical.

Figure 11.5 shows what proportion of the population might actually agree with the garbage tax, given different sample sizes. If the sample were of only 10 people, then as few as 38 percent of city residents might actually agree with the new tax, even though the survey suggests 70 percent will agree; of course, it could also be true that 94 percent of the population might actually agree. Note that as you move down the table, from smaller to larger samples, the margin of error narrows. With a sample of 1000 people, we can be reasonably confident in predicting that the percentage of people in the population agreeing with the garbage tax is between 67 percent and 73 percent—that is, within +/− 3 percent of 70 percent. But what does it mean to be "reasonably confident"?

Level of Confidence

Polling results are frequently reported with the proviso that they are accurate "19 times out of 20." This reflects the sampling dilemma referred to in the introduction to this chapter. Since we have talked with only a subset of the population, we cannot be completely confident that our sample perfectly represents the population. To say that 95 percent of the time (i.e., 19 times out of 20) the sample estimate is within +/− 3 percent of the population value is simply to acknowledge that chance or random error is present in the sample selection process.

CALCULATING SAMPLE SIZE

In theory, efficient sample sizes can be calculated if you know how accurate the results must be as well as how much variation exists in the population. The necessary level of accuracy depends on the kinds of consequences or decisions that are to be based on the research results. If you need results about which you can be supremely confident, you need a larger sample. Understanding how survey results will be used—and knowing the costs required for different sample sizes—makes decisions about accuracy (and hence sample size) relatively straightforward.

Estimating the variability in a population is more difficult. The primary reason for doing a survey is to learn something about a population, so knowledge of variability in the population is often not readily available. There are some methods that can be used, however, to estimate variability:

1. *Ask experts.* You can ask people knowledgeable about a population to estimate rates of variability for key variables.
2. *Use a pilot test.* From a very small, random sample of the population, you can calculate measures of variability to use in determining sample sizes.
3. *Use previous results.* Sometimes the results of earlier research can be used to estimate variability.
4. *Make an educated guess.* As a last resort, estimate the lowest and highest values (i.e., the range) on a key variable and divide this range by four (Henry, 1990: 119).

The estimate of variability can then be used, along with decisions about required accuracy and margin of error, to compute the required sample size. Elementary statistics textbooks provide tables or formulae for precise calculations. We caution, however, that these statistical calculations are based on theoretical, not practical, considerations, and survey researchers emphasize that, in practice, they are often of little help in designing real studies (Fowler, 2002). This is because most social surveys do not have one key variable whose measurement precision can be used to decide on a total sample size. Surveyors typically have a host of "key variables" involving multicausal designs (see Chapter 5).

Political polls done for the news media are an important exception. In this case, one key variable is often of main interest—the percentage of people favouring one or another candidate. For these polls, a statistical formula for sample size is more relevant. Alternatively, as explained earlier, Figure 11.5 provides an easy method of visualizing how samples of different sizes vary in the amount of precision we can safely assume for a sample estimate.

Practical Considerations in Calculating Sample Size

Experienced survey practitioners stress three practical considerations when determining sample size. Samples based on principles of probability are important, but attention must also be paid to response rates, subgroup analysis, and cost.

Response Rates

Often we are unable to obtain answers from all of the units in a sample. In the context of general population surveys, there are three major reasons for this: (1) some people refuse to participate, (2) others can never be contacted (e.g., they are rarely home), and (3) still others are ineligible (e.g., in our student sample, we may contact people who have dropped out and are no longer students). Similar problems of access and institutional agreement frequently limit response rates in surveys of organizations.

For these reasons, initial sample sizes need to be adjusted to account for the fact that not everyone you want to survey will provide usable information. Because of non-contacts and refusals, only four out of every five people you hoped to include in your survey may actually participate. With an 80 percent completion rate, to obtain information from 1000 respondents, you might need an original sample of 1250. You might randomly select 1000 students initially and have a backup pool of 250 others, randomly chosen to use as substitutes for students in the initial pool that you cannot contact. Alternatively, the initial sample could include 1250 students, based on the assumption that there will be an 80 percent response rate, leading to approximately 1000 survey respondents. To still consider this a probability sample, you need to assume that respondents and non-respondents do not systematically differ on key variables (if they do, then you do not have a random sample that is representative of the population). Thankfully, this is an assumption that can be tested.

Subgroup Analysis

Frequently in survey research, we want to examine the relationships between several variables. A sufficiently high number of respondents are needed to do this properly. For example, to assess pay differences between male and female teachers, with a control for possible age differences, a sample of 100 would be inadequate. At best, there would be a minimum of 25 people in each subgroup (e.g., young men, older women), and this is too small a sample from which to calculate reliable estimates of average pay. Finer breakdowns with more variables would be impossible. If you wished to do a separate analysis of pay levels for single-parent teachers, a far larger sample would be needed because the subgroup of single parents would be very small in a sample of 100.

Attention to the kinds of analysis required often alerts researchers to the need for larger samples. City hall may want to know the level of support for a new excess garbage tax among all taxpayers, *and* among taxpayers from different neighbourhoods within the city. You must ensure that each neighbourhood will be sufficiently represented to provide reliable estimates. The size of the subgroup in each neighbourhood will then determine the overall sample size. Figure 11.5 shows error margins for small samples, suggesting that upward of 150 to 200 cases are needed for even modest levels of precision.

Cost

Larger samples cost more money, so budget restrictions may dictate sample sizes. The sample size is the largest single cost factor in any general estimation of survey costs. The number of interviews to be done (salaries, travel, phone bills) or the number of questionnaires to administer (printing, postage, labour) influence cost. The size of your budget may be the factor that limits the size of the sample. Only in Internet surveys is the sample size less influential as far as overall costs are concerned.

EVALUATING SAMPLES

Once a survey has been completed, it is important to evaluate the representativeness of the sample. This requires that you try to match the distributions derived from your sample with known distributions of the population. To return to our sample of students: most schools have some basic data about the student population—the numbers of women and men, the numbers of students at each level and in each program, and so forth. By matching sample distributions with the known distributions for the population, you can estimate how well your sample represents the population on those variables. If 55 percent of your respondents are women, and if women make up 52 percent of the school population, you have done a reasonable job of sampling (at least so far as sex is concerned).

For special populations (e.g., students, workers, organizational members), institutional data are often available to assess the representativeness of a sample. For general community, provincial, or national samples, comparing sample distributions with the most recent census results is a good way of assessing the adequacy of a sample. Such comparisons test the claim that a sample is representative of a population.

Survey research is often judged on the basis of response rates. A basic tenet is that higher response rates assure better quality. However, this only follows if non-respondents and respondents actually differ. Put another way, only if non-respondents are distinctive in comparison to respondents does their exclusion from the survey sample actually lead to bias or distortion in survey results. For example, Keeter et al. (2000) report results from two parallel surveys: one with a response rate of 37 percent and another with a rate of 61 percent. They found very few significant differences between the two surveys on a range of demographic, attitudinal, and behavioural questions (see also Curtin, Presser, and Singer, 2005). As a final example, most political opinion polling conducted just prior to the time of an election does a relatively good job of predicting the outcome, even though response rates to political polls are often low. The reason the survey polls are relatively accurate is that respondents and non-respondents do not differ markedly in their voting preferences.

NON-PROBABILITY SAMPLES

There are occasions when the use of non-probability samples must be used. For example, Hagan and McCarthy (1999) were interested in crime among homeless youth. Finding a sampling frame for this population was impossible. They used non-probability methods to generate a sample of homeless youth they contacted either through service agencies (e.g., youth shelters, counselling centres) or in locations where these youth often congregate (e.g., parks, street corners). For many other groups, finding or constructing sampling frames is difficult or impossible (e.g., illegal immigrants, non-prescription drug users, poachers). This does not mean that surveys are pointless with these groups; it simply means that, although generating probability samples is impossible, systematically asked questions may still provide extremely valuable information. However, you need to be careful about generalizing the results.

SUMMARY

Surveys involve only a proportion of a population. A key attribute of sample surveys is the degree of correspondence between the sample and the population. The design of a good sample frame is a key step in enhancing the representativeness of the sample. The laws of probability provide systematic rules for the selection of a representative sample from the sample frame. Exactly how many sample units to select is a difficult question. Sample size is typically determined on the basis of practical considerations: response rates, cost, and analysis requirements.

A final but essential caveat is in order. Although increasing the sample size reduces the margin of error of estimates, it cannot compensate for poor questions or poor design. Errors caused by faulty questions, vague indicators, or improper interviewing techniques cannot be compensated for by increasing the sample size. Sampling bias and sampling error are only two components of the overall or total design.

EXERCISE 11.1 DEFINING A SAMPLING FRAME

Specify the population to which you want your study results to be generalizable. Define a sampling frame for your study. Explain ways in which the sampling frame may deviate from your population. If possible, suggest steps that might be taken to improve the sampling frame. Comment on the feasibility of undertaking such corrective measures. Suggest what (if any) consequences might follow from remaining differences between your population and your sampling frame.

FURTHER READING

Barnette, V. (1991). *Sample survey principles and methods.* London: Edward Arnold.

Brick, M.J., et al. (1995). Bias in list-assisted telephone samples. *Public Opinion Quarterly, 59*, 218–35.

Groves, R.M., Fowler, F., Couper, M., Lepkowski, J., Singer, E., and Tourangeau, R. (2004). *Survey methodology.* New York: John Wiley.

Philipson, T. (1997). Data markets and the production of surveys. *Review of Economic Studies, 64*, 47–72.

Satin, A., and Shastry, W. (1993). *Survey sampling: A non-mathematical guide.* Ottawa: Statistics Canada.

Chapter 12

SAMPLING: METHODS OF SELECTION

Many different sampling strategies can be used when applying the principles of probability sampling discussed in Chapter 11. Five methods are presented in this chapter: simple random sampling, systematic sampling, stratified sampling, multistage cluster sampling, and sampling by random-digit dialling (RDD). RDD requires specialized administration, and so we discuss it in a separate section at the end of the chapter.

To appreciate how the other sample selection methods work, we will use as an example a hypothetical survey of Canadian businesses. The fictitious survey is designed to examine the social policies of large Canadian businesses—policies such as parental leave, mandatory retirement, leave to hold political office, and sick leave. The research question is whether large foreign-owned and Canadian-owned companies differ in the range of social benefits they offer to employees.

In any sampling problem, establishing a sample frame is a key step. The unit of measurement in the proposed survey is the large business. We could define "large" by the number of employees or by revenue (or some combination thereof). A list of such companies is published frequently in various Canadian business magazines. (In an actual study, the method of composing the list would require careful scrutiny and possible alterations.) For our example, Figure 12.1, we use a ranking of Canada's 500 largest companies, compiled by the *Financial Post*.

FIGURE 12.1 LIST OF LARGE COMPANIES IN CANADA

Sequence Number	Company Name	% Foreign Owned
01	Abbott Laboratories Ltd.	100
02	Acklands Grainger Inc.	100
03	Aecon Group	0
04	Agrium Inc.	0
05	Agropur Cooperative	0
06	Air Canada	0
07	Alimentation Couche-Tarde Inc.	0
08	Allianz Canada	100
09	Apache Canada Ltd.	100
10	AstraZeneca Canada Inc	100
11	ATCO Ltd.	0
12	ATI Technologies Ltd.	0
13	AXA Canada Inc.	100
14	Barrick Gold Corp.	0
15	BASF Canada	100
16	Bayer Inc.	100
17	BC Gas	0
18	Bechtel Canada Co.	100
19	Bell Helicopter Textron Canada Ltd.	100
20	Biovail Corp.	0

For our study, we need policy profiles from a sample of 200 companies. This decision about sample size is based on our survey budget, our estimate of the range of policies in different firms (i.e., sampling variability), and our planned tabular analyses of the data. We also know that even with our best efforts not all companies will respond to our survey, so we elect to sample 250 companies with the aim of obtaining usable data from at least 200 firms.

Figure 12.1 provides an alphabetical listing of the first twenty companies in the top 500 companies (our sampling frame). The sampling frame is the list of 500 companies we alphabetized and numbered sequentially. In order to make the sample work as an illustration, we omitted banks, crown corporations, and companies with only partial foreign ownership. Based on our sample-size decision (i.e., 250), we need to randomly select one out of every two companies.

SIMPLE RANDOM SAMPLING

In this method of sampling, we pick a number at random and then check to see if the selected number matches any of the sequential numbers in our sampling frame. If a match is found, the unit, or company, identified by that number becomes part of the sample. Computers can generate these random numbers, or tables of random numbers (see Figure 12.2) can be used (such

FIGURE 12.2 TABLE OF RANDOM NUMBERS

44	70	48	29	49	89
10	09	73	25	33	90
37	54	20	48	05	02
08	42	26	89	53	04
99	01	90	25	29	11
12	80	79	99	17	71
66	06	57	47	10	01

tables are included in most elementary statistics books). Random number generators are built into most computer data analysis packages (e.g., SPSS—*Statistical Package for the Social Sciences*). The tables simply provide a convenient way of ensuring that probability laws are being followed in generating the sample (i.e., that randomness prevails, with no systematic patterning of numbers).

The random numbers in Figure 12.2 are used to select businesses from among those listed in the sampling frame (Figure 12.1). To simplify this example, consider the first 20 companies listed in the sampling frame (Figure 12.1). Since we want 250 companies out of the full list of 500, then we want a sample of 10 companies from our list of 20. Using the table of random numbers in Figure 12.2, we can randomly select businesses from our frame of 20. We can start anywhere on the table of random numbers, but for ease of presentation we will begin in the upper left corner, and proceed down the column, selecting two-digit numbers. The first number, 44, does not match any number in the sample-frame list. However, the next number, 10, matches AstraZeneca Canada and that company is selected for our sample. The next number, 37, does not match, but number 08 matches Allianz Canada and that company is also selected.

Continuing down column 1 we also select 12 (ATI Technologies). If we then jump to the top of column 2 and continue down we choose 09 (Apache Canada), then 01 (Abbott Laboratories), and then 06 (Air Canada). In column 3 the only match is 20 (Biovail). We would then select 05 (Agropur Cooperative) and 17 (BC Gas) from column 5. The next number is 10, but since AstraZeneca Canada is already included, we ignore this selection. Finally in column 6 we would select 02 (Acklands Grainger), thus giving us the 10 randomly selected companies needed.

What makes this a *simple* random sample is the fact that each company has an equal chance of being included in the sample. In other words, the probability of any one company being selected is 1 in 2, or 50 percent. One issue complicates simple random sampling: once an element has been selected,

FIGURE 12.3 PUTTING STUDENTS BACK INTO THE HAT: SAMPLING WITH REPLACEMENT

should it be eligible for re-selection? For example, the number 10 was encountered twice in our list of random numbers (it appeared in both columns 1 and 5). Under these conditions, should we have selected AstraZeneca twice, or were we justified in ignoring the second selection of the number 10? There are two options from which surveyors must choose when faced with this question: sampling with replacement or sampling without replacement.

Sampling *with replacement* means that any element may be included more than once (i.e., it gets put back, or replaced, so that it may be chosen again). Thus, if the company AstraZeneca were included twice in our sample, it would have double the weight in any results. It seems intuitively wrong to allow one element to have double the influence on the results, and frequently survey researchers choose not to sample this way. However, especially for small populations, sampling with replacement is sometimes the preferred strategy as it ensures that each possible sample has the same probability of being selected.

The alternative is to sample *without replacement,* meaning that once an element has been selected, it cannot be chosen again. This strategy technically violates one of the principles of probability theory (i.e., that all possible samples should be eligible) because once a company has been selected it cannot be chosen again. With large populations, the difference in the two methods is negligible, and thus sampling without replacement is the typical practice in these cases.

SYSTEMATIC RANDOM SAMPLING

A more efficient way of obtaining a sample is to use a systematic sampling method. One drawback with simple random sampling is that it is laborious, since you must continually move between the randomly selected numbers and the sampling frame to identify matched numbers (unless you can use a computer to search and match). An easier method is to use the sample-frame list, decide how many elements on the list need to be selected (i.e., the total sample size), then choose every kth element on the list, where k is determined by dividing the number of elements in the sample-frame list by the sample size. So, for our list of 20 companies (sampling frame), we divide by the sample size of 10, which results in $k = 2$.

Once again, from our list of 20 companies we need to select 10 firms. To avoid bias, we must choose a random starting point, then select every second company. Our random start must be either the first or second company on our list, since we want every second company. We can simply flip a coin to choose a starting point or use our table of random numbers again. Since you cannot see our coin flip, we will use the more laborious method with the random numbers in Figure 12.2. Now we focus on single digits and proceed down column 1 until we find either a 1 or a 2. After the number 4, the next single digit number is a 1, which gives us our randomly chosen starting point. Since we are to select every second element, starting with element 1, we would select the first, third, and fifth through to nineteenth companies on the list. Note that we are sampling without replacement, since no element can possibly be selected twice.

Systematic sampling relies on a *sampling interval* (called k above) and a *random start*. The sampling interval (or skip interval) is the numerical distance between two selected elements. The sampling interval depends on the size of the population and the size of the sample. Calculate the sampling interval by dividing the population size by the sample size. In the example, divide 20 (population size) by 10 (sample size) to obtain k, the sampling interval (here, 2).

The starting point is a randomly selected number between 1 and the sampling interval. In our example, we had to choose randomly between 1 and 2. If our sampling interval had been 10, we would have chosen a random start between 1 and 10. An important complication occurs when the calculated sampling interval turns out to be a fraction. If this happens, always round the number down to the nearest lower whole number. For example, with a sample frame containing 201 elements and a required sample size of 50, the sampling interval is 4 (201 ÷ 50 = 4.02, rounded to 4). Note one last twist. If a random start of 1 is selected, the following elements should then be selected for the sample: 1, 5, 9, 13 . . . 201. This would result in 51 selected elements (any other random start would give only 50 selected elements). All 51 elements can be included in the sample, or, if exactly 50 is essential, one selected element should be randomly discarded.

A possible complication with systematic random sampling comes with a list that has some repetition or cyclical quality (technically referred to as *periodicity*). For example, consider addresses in a large apartment building. There might be 12 units per floor, and the units ending with 12 (i.e., 112, 212, etc.) might all have magnificent views and therefore cost more to rent. A problem occurs if a systematic sample is used to choose one unit per floor. Choosing every twelfth unit, with a random start of 12, would always result in the more expensive units being selected; with any other random start, an expensive unit would never be included—either way, the sample would not be representative of the units in the building.

There are, however, other occasions when systematic samples are advantageous. If we could easily rearrange our list of companies from the largest to the smallest, a systematic random sample would guarantee the proper proportion of large and small companies in our sample. In contrast, if we used a simple random sample, large or small companies could be over- or underrepresented just by chance. By choosing a systematic sample, selected from a list of companies ordered by size, we are guaranteed that our sample contains a proportion of large and small companies nearly equal to their proportion in the population.

STRATIFIED SAMPLING

When the population from which a sample is being taken is heterogeneous, one way to reduce the effects of heterogeneity is to organize the population into subgroups or strata, as long as the variable or variables by which the strata are organized are related to variables in the research question. The representativeness of a sample of university students for a survey on choice of academic discipline would be enhanced if the population were stratified into male and female.

Gender stratification would produce a better sample because it would more accurately reflect the population distribution of women and men. We also know that women and men enroll in different academic streams (e.g., more women in nursing, more men in engineering). If we constructed two sampling frames, one for women and one for men, we could then choose a random sample from each frame to reflect the population distribution of women and men at a particular school (which might be 55 percent women and 45 percent men). Not only would our stratified sample reflect the sex distribution better than would a simple random sample, but it would also more accurately reflect the population distribution among academic disciplines. This latter result would occur because sex (our stratifying variable) is related to course choice (a key research variable).

FIGURE 12.4 STRATIFYING THE POPULATION BY SEX

Stratified sampling can also be illustrated using our example of large companies in Canada. The corporation sample is effectively stratified by arranging the list of companies from largest to smallest. As opposed to the initial alphabetical arrangement, the list now has strata or subgroups, in that large companies are listed beside other large companies and small companies are listed beside other small companies. This new way of listing improves the precision of the sample by guaranteeing proportional representation by company size (assuming we use systematic random sampling). Similarly, a list of employee e-mail addresses sorted by a store's location could be systematically sampled so that the number of employees selected per location was equal to the proportion of employees working at each location.

We might go one step further in the stratifying process with our list in Figure 12.1. Aside from the criterion of size, it is worth noting that 10 of the firms listed are owned by foreigners, while the other 10 companies are owned by Canadians. Company social policies might vary by type of ownership—the research question with which we opened this chapter—and having a sample that contains a representative number of both foreign- and Canadian-owned companies helps in testing that question.

One way of doing this is to divide the list into companies that are owned by foreigners and those that are owned by Canadians. We can then choose a random sample from each list and thus be assured that we have the proper

FIGURE 12.5 STRATIFIED SAMPLING OF LARGE CANADIAN COMPANIES

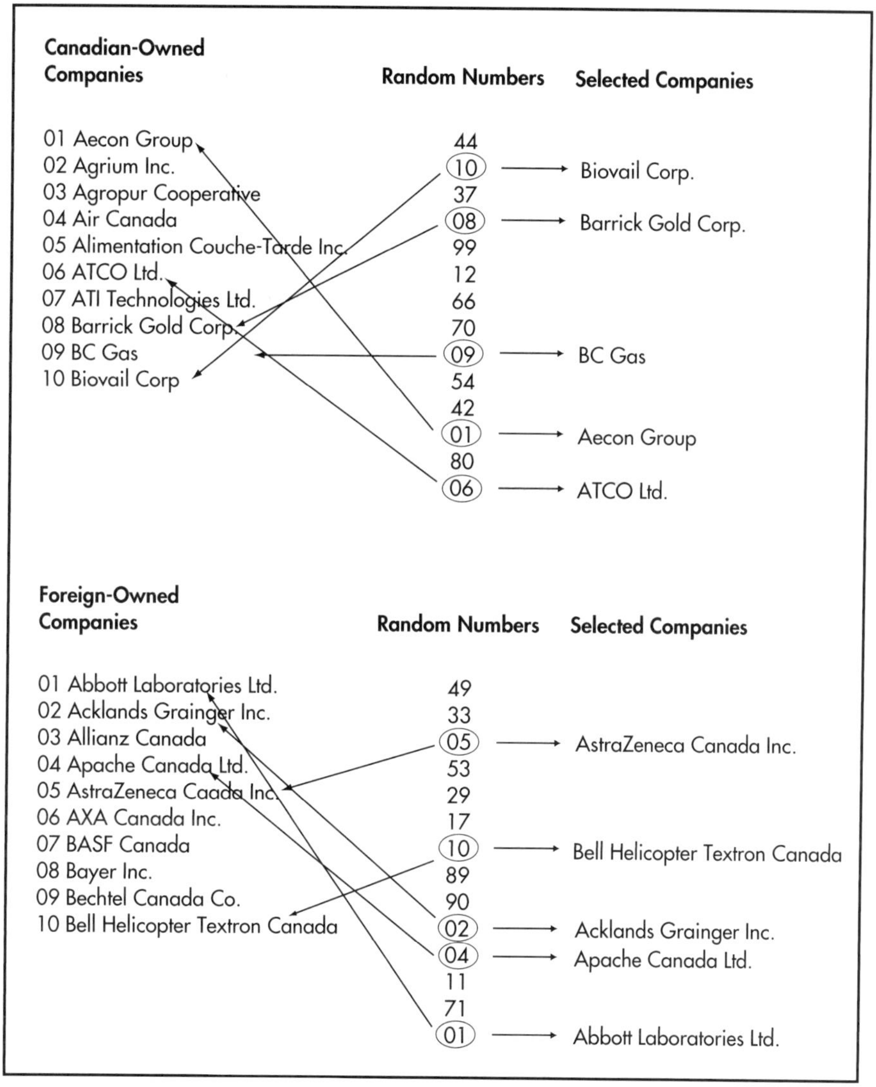

balance of Canadian and foreign-owned firms. Systematically choosing every second company from each of the two lists gives us five firms of each type. Alternatively, we could also use the random numbers from the table in Figure 12.2 to select companies for the sample from each list, as illustrated in Figure 12.5. When we chose a simple random sample (above), we actually selected five Canadian companies and five foreign firms. The selection of an even number of foreign and Canadian firms occurred by chance, in a manner analogous to obtaining five heads when flipping a coin ten times. It was simply fortuitous; we could not foresee that it would happen. By using a sample stratified by ownership type, we can guarantee that the sample more accurately reflects the population.

Information about the population is needed before we can proceed with stratified sampling. Obviously, we can only stratify the sample using information that we know. If we had not had a list that identified company ownership, we could not have stratified by that variable. Another example of the same issue involves taking a sample of university or college students, where it might make sense to stratify by field of study in order to ensure that people from all fields are included. We could only do this, however, if we had a list of students whose field of study was identified. Without such information, we could not stratify the sample using that variable.

Stratified sampling can also be used to ensure that samples contain enough cases of particular sub-populations. For example, using the company example cited above, we might want to undertake a comparison of new and old companies. There are not likely to be many new companies among the top 500 firms. We need a way to ensure that a sufficiently large group of these newest companies appears in our sample.

We could do this by assigning *unequal* probabilities of selection to the elements in the sampling frame. Again, we might create two lists: a list of new companies and a longer list of older companies. To ensure that we have enough young firms in our sample, we could choose two out of every three new companies and one of every three old companies. With only a few new companies in the population, this is the only way of guaranteeing that enough new firms appear in the sample to allow for meaningful analysis; of course, when making general conclusions based on our results, we would have to correct for this unequal sampling.

A second example might be helpful. Assume that in our student sample we are interested in comparing foreign students and Canadian students. Proportionately, there are far fewer foreign students registered at colleges and universities. Rather than giving all students an equal chance of being in our sample, we might deliberately provide foreign students with a greater chance of being selected. If we could stratify our sampling frame into Canadian and foreign students, we could over-sample foreign students to ensure a sufficient sample size. We would have to take this unequal sampling into account in presenting results. However, *disproportionate stratified sampling* is a powerful method of sample selection that can be used very effectively if there are subgroups in a population that require over-sampling in order to ensure a sufficient number of respondents in all subgroups.

The variables by which we stratify the sample are either variables in the research question or closely related to variables in the research question. For a research question that is focused on job satisfaction, stratifying a population by age of the neighbourhood in which people live does not improve the representativeness of the sample, unless the age of the neighbourhood is related to job satisfaction. Similarly, in the sample of Canadian corporations, stratifying by corporate colours (if that were possible) is not likely to improve the representativeness of our sample with respect to company social policies. Only variables relevant to the survey problem are relevant to the generation of a stratified sample (e.g., whether or not the company is unionized would be useful in this case).

MULTISTAGE CLUSTER SAMPLING

Although the list of the *Financial Post*'s 500 largest corporations is a convenient sampling frame, it represents only a small fraction of all the firms in Canada. If we wanted to take a sample from *all* businesses in the country, the sampling frame would include several hundred thousand companies, most of which would be very small businesses. Mailed questionnaires could be used as the data-collection method, but it might be more effective to use interviewers, who could track down and encourage busy people to participate in the survey. However, to send interviewers across the country would be time-consuming and extremely costly. A better sampling strategy might be to target firms in specific areas, and draw samples from only these delimited places.

First, we could randomly select a sample of Canadian cities, then randomly select neighbourhoods within each city, and finally randomly choose firms within each of the selected areas. Our sampling strategy would evolve in stages (cities, areas, firms), and our sample would be clustered because only firms in the selected cities, in randomly chosen neighbourhoods, would be included.

The advantage of this *cluster sampling* is that interviewers need only visit selected cities and, within those cities, specific areas. The disadvantage is that businesses in a particular city, or area of a city, tend to be similar. If you hit on the waterfront area of a coastal city in the first selection round, you might get an abundance of small-boat building or fish-processing firms. Alternatively, a second-round selection of a city district might include either a neighbourhood with high-priced retail stores or many second-hand retail shops. Clusters will tend to be internally homogeneous. When using any form of cluster sampling, the rule is that, whenever possible, the surveyor should choose more clusters with fewer elements per cluster.

Multistage cluster sampling, also known as *area probability sampling*, is frequently used when national or provincial face-to-face household interviews are done. A popular method of sampling in these cases is to first select a sample of census tracts, then select census enumeration areas (EAs), then streets in

FIGURE 12.6 USING CENSUS ENUMERATION AREAS (EAs)

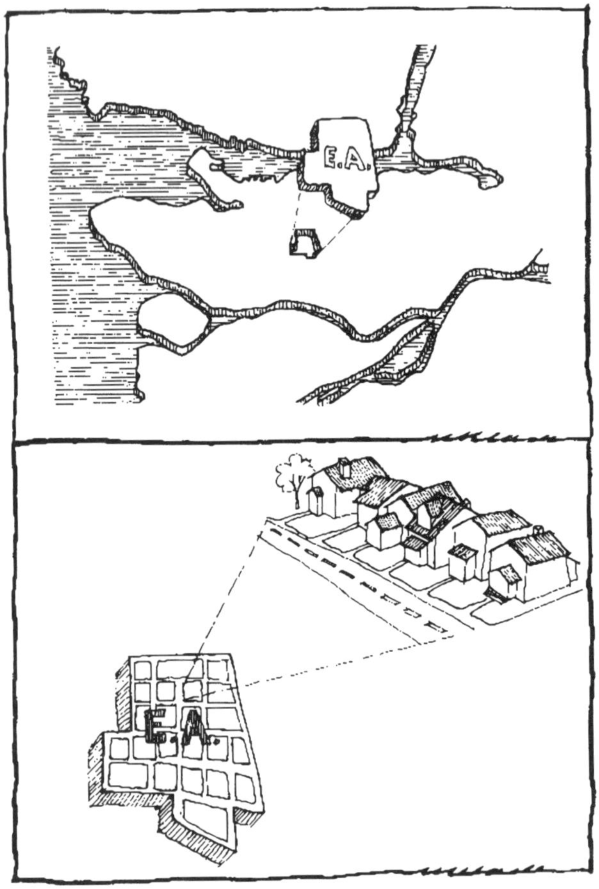

selected EAs, and finally list the household addresses for the selected streets (see Figure 12.6). Statistics Canada can be consulted for lists of census tracts and EAs, while city directories provide streets and addresses (although directory information requires careful checking for missing household addresses). Finally, a sample of households is chosen, and an interviewer is sent to each address. The key to area probability sampling is that, at each stage, a list is established (the sampling frame) and a random sample is drawn from the list.

After a sample of households is drawn, a method of choosing which household member to interview is required. A typical strategy is to use a set of household selection grids (see Figure 12.7). At each randomly selected household, the interviewer asks two questions: how many people in the household are over 18, and how many of these are men (or how many of these are women?). If there is more than one person living at a selected address, the interviewer is able

FIGURE 12.7 A HOUSEHOLD SELECTION GRID

2. How many of these people are men?	1. How many people 18 or over live at this address?	2	3	4 or more
	0	Youngest woman	Oldest woman	Second oldest woman
	1	Woman	Man	Youngest woman
	2	Youngest man	Youngest man	Oldest man
	3		Middle man	Oldest man

to choose which person in the household to interview by using a grid. There are several different grids, in which different household members are assigned to cells. For example, in one grid the person chosen in a household of two people over 18 where one is a man would be the woman, as in Figure 12.7, while in another it would be the man. Grids are randomly assigned to addresses, thereby ensuring that the person is chosen randomly for different addresses. Alternatively, research firms often ask to interview the eligible household member (e.g., over 18) with the most recent birthday (asking for the person with the next birthday is more confusing; see also Gaziano, 2005).

Area probability samples are usually stratified. Variables that have been used to stratify area probability samples include the size of the area (e.g., EA, streets) or typical characteristics of households in an area (e.g., socioeconomic status, ethnic mix).

RANDOM-DIGIT DIALLING

A sizeable and increasing percentage of all surveys are done by telephone. This is due, in large part, to the emergence of random-digit dialling (RDD) as a sampling method. For several reasons, samples that are based on telephone directories have always been problematic. For example, many people (especially married women) are not listed by name. As well, many people move during the year and their names no longer coincide with the old addresses. Even as a sampling frame for household addresses, directories are biased because a growing proportion of homes have unlisted numbers, and some homes (about 2 percent in Canada) have no phones.

RDD avoids these problems (except for homes without phones) while taking advantage of the fact that the vast majority of households have working telephone numbers, even if a perfectly accurate listing of those numbers is not publicly available. Telephone companies assign numbers in groups. The number 416-445-XXXX has first an area code (416 being one of the area codes used for Toronto) and then an exchange number (445). Area codes and working exchange numbers can be obtained from telephone companies or sampled from telephone directories. One method of dialling numbers randomly is to assign the next four digits of the telephone number randomly. The problem with this strategy is that many numbers would either be out of service or belong to businesses rather than households. To increase the number of working household numbers reached, you could determine the first eight numbers (e.g., 416-445-79XX) and simply dial the last two digits randomly. With RDD, no explicit sampling frame exists, but implicitly all households with working telephone numbers are eligible for inclusion.

So long as the necessary survey information can be obtained over the telephone, RDD is a useful method of sampling large populations. The technique can also be used, in conjunction with screening questions, to identify specialized populations. To reach a random sample of households with teenagers, RDD can be used to contact households, and a couple of quick screening questions can ascertain whether this household qualifies (i.e., has teenagers). Telephone numbers are typically assigned to households and not individuals; therefore, once an RDD connection has been made, the grid selection method described above (or some substitute) is used to randomly select respondents from among all eligible household members.

RDD does not work well, however, when you wish to sample people in small geographic areas that do not match the boundaries of telephone exchanges. In cases like this, a significant number of phone calls may be made to households outside the small geographical area. Also, caller ID and increased telemarketing have been linked to falling response rates on telephone interviews, because respondents are more likely to answer phone calls only from numbers they recognize (see Curtin, Presser, and Singer, 2005).

SUMMARY

Several methods have been devised for choosing samples based on the rules of probability. In simple random sampling, each element in the sampling frame has an equal chance of selection. Systematic sampling makes use of a sample-frame list and requires the selection of a fixed sequence of elements (e.g., every fifth element) from the sampling frame. Stratified sampling requires some means of subdividing the sampling frame prior to the random selection of elements. This method, especially if used with unequal sampling probabilities for different elements, is excellent for ensuring a sufficient number of cases in different sub-populations. Stratified sampling also improves sample accuracy by ensuring that a representative proportion of elements is selected on the stratifying variables. Multistage cluster (or area) sampling uses a series of stages generally involving the random selection of smaller and smaller clusters (or areas) at each stage until, at the final stage, a specific respondent is selected. RDD uses principles of probability to select telephone numbers for dialling, which generates randomness among respondents on the basis of which households are called. As well as discussing specific details about how to use each selection method, the chapter also reviewed some of the drawbacks associated with each method.

EXERCISE 12.1 GENERATING A SAMPLE OF RESPONDENTS

Choose a method for selecting a probability sample for your study. Discuss the advantages and disadvantages of this method as it applies to the population and sample for your study. Using the sampling frame you developed in Chapter 11, show *in detail* exactly how you would go about choosing a specific sample. You must provide sufficient detail so that someone else could follow your outline to actually generate a list of respondents to survey.

FURTHER READING

Henry, G.T. (1990). *Practical sampling.* Newbury Park, CA: Sage.

Lavrakas, P.J. (1993). *Telephone survey methods: Sampling, selection, supervision* (2nd ed.). Newbury Park, CA: Sage.

Tryfos, P. (1996). *Sampling methods for applied research.* New York: John Wiley.

Chapter 13

PROCESSING AND REPORTING THE DATA

As we have noted earlier, all survey data is entered into a computer, either automatically or manually. This allows survey researchers to take advantage of powerful software packages (both quantitative and qualitative) to summarize the vast amount of data typically collected. Once the data is summarized, it can be used in reports tailored to specific audiences. In this chapter we focus on the following questions: how do you transfer the answers of respondents to electronic databases, and how do you describe the basic results of a survey research project?

PROCESSING THE DATA

With an Internet survey, the computer can be programmed to automatically store responses in a database. The same happens with computer-assisted interviewing. The process is relatively simple, except that you have to understand how to organize the data—an issue that will become clear as we examine some of the complications involved in coding data. With paper-based, self-administered questionnaires or interviews[1] where computers are not used, the creation of

[1] For the sake of brevity, the term "questionnaire" will refer to both questionnaires and interview schedules.

FIGURE 13.1 FREQUENCY TABLE

ANXIOUS	Anxious when applying for personal line of credit				
Value Label	**Value**	**Frequency**	**Total Percent**	**Valid Percent**	**Cum. Percent**
STRONGLY DISAGREE	1	110	16.9	17.2	17.2
DISAGREE	2	149	22.9	23.2	40.4
NEUTRAL	3	167	25.7	26.1	66.5
AGREE	4	150	23.0	23.4	89.9
STRONGLY AGREE	5	65	10.0	10.1	100.0
NO ANSWER	9	10	1.5	MISSING	
Total		651	100.0	100.0	
Valid Cases 641	Missing Cases 10				

electronic files requires a careful entering of data into a well-conceived database. Below we explain the procedures followed when setting up databases and preparing data for electronic analysis.

Consider the following example. A financial institution wishing to relieve customer anxiety during the loan-application process employs a market research firm to interview loan applicants. During the interview, customers are presented with a set of categories describing anxiety and are asked which single category best fits their response to the statement "I am anxious when applying for a personal line of credit."

The results, summarized for all respondents, are displayed in Figure 13.1. (One objective in any analysis of data is to produce such tables for all variables in a survey.) All the information necessary to read and understand the table is provided. The reader knows the name of the variable—anxious. In cases where the name of the variable is not sufficiently informative for readers, an extended variable name or variable label is given. Values and their labels also appear in the table. If some respondents did not answer the question or the question did not apply to them, the number of missing cases is listed.

This probably looks and sounds simple: ask a question, record the answers in a data file or database, and then present results. For the most part it is straightforward, but there are some important tricks of the trade that will be helpful.

CODING THE DATA

Coding involves transferring the survey information to computer data files. The *codes* are the values used to represent different responses, the *coder* is the person who transfers the data, and the *codebook* is the written set of coding instructions provided to the coder.

What do we mean by codes? For statistical analysis, we typically use numeric representations, or *codes,* for each value of every variable. For example, if the respondent answers "female" to the question "Are you female or male?" we usually choose a number such as "0" or "1" to represent female respondents. A different number, typically either "0", "1", or "2", is then assigned to male respondents. In simple presentations of frequency or percentage of respondents who are in a set of categories, some analysts prefer to code and present the name of the category, e.g., female or male, rather than using a numeric representation. However, using text rather than numbers precludes some types of statistical analysis of the data. Using the example of sex, Figure 13.2 presents the various terms used in setting up the file for coding and for subsequent statistical analysis. In this case, "Sex" is defined as the one-word *variable name;* "Sex of Respondent" is the label or *extended variable name;* "Female" and "Male" are the possible responses or *value labels;* and "1" and "2" are the numerical codes or *values* assigned to represent the possible responses.

In the case of computer-assisted questionnaires or interviews, the coding scheme is developed as part of the questionnaire. Responses to questions are directly and automatically entered into the computer database. Variable names, value labels, codes, and codebooks are still part of the process, but the data entry is done directly by either the interviewer or the respondent when the questions are being answered.

In all cases, the process of going from survey responses to computer-analyzable data follows the same process, even though the steps may be handled differently. The process the researcher follows is dependent on

1. whether the data gathered in the survey are recorded as close- or open-ended answers,
2. what software program is being used to enter the data, and
3. whether the codes are contained in the questionnaire.

FIGURE 13.2 STANDARD TERMS USED IN STATISTICAL PROGRAMS

Variable Name: Sex
Variable Label: Sex of Respondent
Value Labels: Female, Male
Values: 1, 2

CODES, CODERS, AND CODING

Close-Ended Questions

Coding is simple with close-ended questions. If the questionnaire has the numerical codes listed on it (see Figure 13.3), coding simply involves transferring these numbers to the computer using a personal computer program such as a spreadsheet program, a database program, or a data-entry program from a statistical software package. Most personal computer statistical programs will recognize data from any of the above types of programs.

The researcher usually assigns numeric values to represent values of the variables, such as "1" for female and "2" for male, even when the numeric values are used merely to differentiate categories. This is because most computerized statistical programs used to analyze the data are restricted to manipulating numeric values. The codes to be used are either defined on the questionnaire (for close-ended answers) or a set of instructions (codebook) is provided to explain to coders the number that represents the different responses for each question.

In coding close-ended questions, a key principle for researchers to follow is to directly code the values contained in the questionnaire. For example, if individual years of schooling are recorded on the questionnaire, individual years of schooling should be coded rather than having the coders group the responses into categories or groupings (e.g., 1 to 5 years, 6 to 8 years, 9 to 11 years, etc.). There is an increase in the likelihood of coding errors occurring when coders transform values. If values need to be transformed, the process can be systematically accomplished with less error by a computer program. Not transforming values at the coding stage allows the researcher to keep the original values and permits the testing of different classification groupings (e.g., 1 to 5 years, or 1 to 8 years) at the data-analysis stage.

Open-Ended Questions

Some types of open-ended questions have the same structure of answers as the fixed-response types, such as age or country of birth. In these cases, all the possible answers are known by the researcher prior to asking the question, and

FIGURE 13.3 USING CODES CONTAINED IN QUESTIONNAIRES

What sex are you? *(Please check the appropriate category)*

Female ❒ 1
Male ❒ 2

appropriate codes can be established before the data are even collected. However, for many open-ended questions, particularly in exploratory research, the range of answers is not known in advance. If the coding task was simply to transfer all the different responses to the computer data file, the only problem would be one of transcribing the information correctly. However, the researcher is looking to define summary categories in which the open-ended responses can be placed. The responses to these types of questions are difficult to code reliably (see Chapter 8). Consider, for example, the following:

Question: What was the single best thing about attending college or university?

Responses given:

1. I had a very active social life.
2. Great parties.
3. I made contacts with people I will cherish my whole life.
4. I learned how to communicate and collaborate.
5. I learned how to play bridge.
6. I learned how to think.
7. I learned how to discipline myself.
8. I learned how to organize my time.

How do you handle 800 or even 300 such answers? The researcher cannot leave the categorization of the answers up to the coders, interviewers, or respondents. They usually do not have a stake in the survey, or the knowledge or experience to group responses appropriately. They are each likely to make different coding decisions as well.

Categorizing Open-Ended Responses

Categorizing open-ended responses involves the following steps:

1. Prior to coding the other items in the questionnaire, each open-ended answer, along with the questionnaire identification number, is entered into a computer database program.
2. The researcher sorts through all the responses and makes decisions about which answers can be combined or grouped with other answers. In essence, a set of response categories is created, and each open-ended response is classified into the appropriate response category. Software packages for handling so-called qualitative data can be used for assistance in this process of creating categories.
3. The categories are named and given numerical codes representing the category names. These codes are then transferred back to the questionnaire

beside the open-ended question. It is usually better to use red ink or some other technique to highlight the information, so that the coder can recognize the code quickly.

4. The open-ended questions are now set up for entry into a computerized data file.

For example, based on the eight open-ended responses above you might use four categories: academics (4 & 6), social activities (1, 2, & 5), personal development (7 & 8), and other (3). There is some arbitrariness with the classification scheme you create, since you are interpreting the meaning of what others said or wrote without knowing exactly what they meant. As discussed in Chapter 6, this is an issue with any measurement process—an issue that occurs using any social research methodology.

Sometimes researchers use open-ended responses as illustrative quotations to enrich statistical reports. Such quotations are a good way of showing readers how survey respondents answered specific questions.

Missing Values

We do not always have information from all respondents for a specific variable. In such cases, we need to provide a code value that signals the information is missing. This value should be outside the range of meaningful values. For a variable with less than 10 values that can be covered by the integers 0 through 8, where 9 is not a valid value, "9" can be assigned as the missing value; for variables having 10 or more but less than 99 values, "99" can be the missing value. Consistency in the values assigned to represent missing data makes coding and analysis much easier; thus, when appropriate, always use "9", "99", or "999" rather than sometimes using "0", other times "99", and occasionally "44". Most computerized statistical programs allow us to define missing data values, often with distinct non-numeric symbols which can then be excluded from the analysis.

Alternatively, you may have missing data that you want to analyze separately. For example, on some questions survey participants might respond "don't know," while for other respondents the question may not be applicable. Keeping these separate may be important. If it is, then you need to plan for this with separate codes for responses such as "don't know," "refused," and "not applicable."

Codebook Construction

When constructing a codebook, the researcher must develop a set of rules and procedures to follow when transferring responses into a computer data file. Each questionnaire receives an identification number (ID) to differentiate it from other questionnaires. In the case of errors in the computerized data set, the ID allows the researcher to check the original questionnaire for the correct code. The responses to each question (variable) will be coded by assigning a numeric value for each possible response, with all the responses making up a *data line*. In turn, all the data lines (at least one for each respondent) will make up a *data set* or *data matrix*.

A data line could look like "01 2 23 1 0 10 6" or like "0122310106", with each number representing the value of a variable, as in "01" (ID), "2" (male), "23" (age), "1" (voted in 2005 election), and so on. While blank spaces between values improve readability for the surveyor's coding and editing, they are not necessary if a fixed format is used for coding.

Fixed Format

The computer must be instructed on exactly how to read the data line. Using as an example the data line containing the blank spaces, the computer is instructed to read columns 1 and 2 together as one number (i.e., "01" in the case above); to skip column 3 (i.e., the blank) and read column 4 as one number; to skip column 5 and read columns 6 and 7 together, and so on. The number of digits to be read as one number is referred to as the *field width,* while the left-hand column where the computer starts reading is referred to as the *field location*. Note that when the value "1" is entered into a field two columns wide, it is written as "01" or " 1", with the space preceding the digit "1" representing a blank (this is known as right justification).

The above process is referred to as *fixed format* because the location of each variable in the data line is fixed. For each variable, then, the computer reads the number from a fixed location in each data line to determine its values. A data set would thus appear as follows:

	variables across the top
units	01 2 23 1 ....,
or	02 2 08 0 ....,
cases	03 1 19 1 ....,
down the	04 1 20 1 ....,
side	05 1 06 0 ....,

The information can either be transferred directly from the questionnaire to a computer file or transcribed onto a coding sheet (or with a Web-based survey, the computer can be instructed, using this codebook logic, to construct a database automatically as survey participants respond). When coders do not enter the data directly into the computer (for whatever reason), the data is first coded from the questionnaires to coding sheets. Data-entry operators then extract information from the coding sheets and enter it into the computer. Commercial firms that provide data-entry service verify the data by entering the information twice and then having the computer check that the two sets of numbers are the same.

In contemporary database programs, the values for a variable are entered into a field. The program automatically pre-sets the width of the field, but these widths can be changed individually or collectively. Allowed values can be defined (e.g., "0" or "1"), or value ranges can be defined (e.g., 1–99). If the entry of a value that is not allowed or falls outside the range is attempted, an error message asking for an appropriate value to be entered comes up. As with the coding of paper questionnaires, value labels can also be defined. Errors can occur individually during the data-entry process, or systematic errors can be introduced if an error in program instructions occurs in setting up the data-entry or data-transformation process. It is imperative that this entry and transformation programming be independently checked by another programmer.

A completed codebook should contain the following information:

- *Question number (or identification).* This makes it clear to the coder where the information is located.
- *Variable name.* Use short names starting with an alphabetic character. Each variable name must be unique. Most social science statistical computer programs use these variable names when commands are issued (e.g., age, occupation).
- *Variable label.* These are longer character strings clarifying the precise meaning of the shorter variable name (e.g., age of respondent, occupation of respondent).
- *Value label.* This identifies each category or value of a variable (e.g., the variable "sex" has two categories:—"female" and "male").
- *Variable value.* A number that stands for or represents each value a variable takes on—for example: "1" (female), "2" (male): or "01" (one-year-old), "02" (two-year-old).
- *Field width.* This identifies the number of digits the values of the variable can take on. If the variable has three possible values, a single-digit field is sufficient.
- *Field location or column location.* This identifies the column in which the values for a particular variable start. It is the extreme left column in a multiple-column field. In most contemporary computer databases, this is not important. Only the sequence of the variables is presented.

FIGURE 13.4 PAGE 1 OF A CODEBOOK

Westview Community Centre
Codebook

Quest. No.	Variable Name	Variable Label	Value Label	Value	Column Location	Field Width
1	ID	Identification #	Person 1	01	1	2
			Person 2	02		
			. . .	. . .		
2	Sex	Sex of Respondent	Female	1	3	1
			Male	2		
			Missing	9		
3	Age	Age of Respondent	1 yr.	01	4	2
			2 yrs.	02		
			. . .	. . .		
			73 yrs.	73		
			Missing	99		
4	Vote01	Vote in 2001 Election	No	0	6	1
			Yes	1		
			Refused	9		

The format of a codebook could look like the one presented in Figure 13.4. Even with data transferred to a database from a Web-based survey, it is imperative to create a codebook. This is necessary because in the future others may want access to the data, and the only way to correctly interpret the data is to refer to a codebook.

Free Format

Figure 13.4 exemplifies a fixed format. Data that are entered in *free format* have the variable values entered in a consistent sequence for all cases, but the specific field location is not fixed. Each code in this case is separated by a space or a comma. Using our previous example, data presented in free format would appear as follows:

```
01 2 23 1 ....    or    01, 2, 23, 1, ....
02 2 8 0 ....     or    02, 2, 8, 0, ....
03 1 19 1 ....    or    03, 1, 19, 1, ....
04 1 20 1 ....    or    04, 1, 20, 1, ....
05 1 6 0 ....     or    05, 1, 6, 0, ....
```

FIGURE 13.5 CODING A SCALE

Quest. No.	Variable Name	Variable Label	Value Label	Value	Column Location	Field Width
16.1	Item 1	Politics & gov't complex	Strongly disagree	1	43	1
			Disagree	2		
			Neutral	3		
			Agree	4		
			Strongly agree	5		
16.2	Item 2	Political parties big	Strongly disagree	1	44	1
			Disagree	2		
			Neutral	3		
			Agree	4		
			Strongly agree	5		
16.3	Item 3	I can help change minds of public officials	Strongly disagree	1	45	1
			Disagree	2		
			Neutral	3		
			Agree	4		
			Strongly agree	5		

In the case of free format, the codebook would require neither the field location nor the field width. Fixed format has been the standard type of data definition. Computer data-entry programs, which are used increasingly, automatically define field locations although they still require details on field width.

Coding Scales

In coding scales, such as Blake's (1985) political efficacy scale presented in Chapter 6, each scale item is defined as a separate variable with the responses (e.g., strongly agree, agree, etc.) coded as the value. Each value is coded as marked on a respondent's questionnaire. Remember, however, that given how the question was worded, the values for item 3 in the political efficacy scale meant the reverse of the values of other items. Figure 13.5 shows the codebook for the first three items of the political efficacy scale. The data are entered into the computer according to which number each respondent circled. *After* data entry, the researcher uses the computer to re-code items as necessary and to determine an overall summary scale score for each respondent (e.g., by summing values across all relevant variables).

Coding Multiple Responses

As Figure 13.6 illustrates, some questions require asking the respondents to pick a number of items from a list (e.g., of sports, hobbies, or organizations in

FIGURE 13.6 MULTIPLE-RESPONSE QUESTION

How did you find out about our product?
(Check as many as apply)

from a newspaper . . . . ❑
from a magazine . . . . ❑
from television . . . . ❑
from a friend . . . . ❑

FIGURE 13.7 CODING FOR MULTIPLE RESPONSES

Quest. No.	Variable Name	Variable Label	Value Label	Value	Column Location	Field Width
18	newspaper	found out about product from a newspaper	no yes	0 1	63	1
	magazine	found out about product from a magazine	no yes	0 1	64	1
	television	found out about product from television	no yes	0 1	65	1
	friend	found out about product from a friend	no yes	0 1	66	1

which they might be involved). When the respondent checks an item, she or he is indicating participation (or non-participation if that same item is skipped).

The most effective coding strategy is to define each item on the list as a variable, with the variable values being defined as "yes" or "no" (see Figure 13.7). With this coding, the computer can be used to generate any or all multiple-response combinations of interest to the researcher.

ERROR CHECKING

The creation of a good survey requires considerable time, energy, and financial investment to gain accurate, useful data. In order to maintain a high quality of data, checking procedures in the coding process must be established and followed. In essence, checking procedures are the quality control mechanisms of survey researchers, designed to ensure that the data gathered are the same data that are coded, entered into computer-readable form, and subsequently analyzed.

It is part of the researcher's task to establish procedures for error checking at each step in the data-transformation process. For instance, if the process includes the categorizing of open-ended data, as well as the transferring of values from the questionnaire to data-entry forms, and from data-entry forms to computer data files, error checks need to be established at each stage in the process. An error-checking procedure that is used when transferring the values from the questionnaire to data-entry forms involves having a second coder re-code the same information and then verifying both entries to determine if identical codes were entered.

Some computer programs are able to compare two files to see if they are identical and, if not, where they differ. Error rates can then be compiled for each question that has been coded to determine if the problem exists only for certain questions. Similar to production quality control in manufacturing, standards are set that require the re-coding of all questions that exceed the error rates. Quality control standards vary by researcher and research problem; some problems require more accuracy, while some researchers may be less concerned about small, random errors. A standard that the authors have used is 2 percent for open-ended questions and 1 percent for close-ended questions.

A routine part of error-checking procedures is to draw a random sample of questionnaires for re-coding for the purpose of quality control analysis. It is recommended that random error checking be continually performed throughout the coding process to ensure the early identification of problems. Potential errors can be further reduced with the use of a data-entry program or procedure. As stated earlier, most data-entry programs can be set to accept only certain values. In the case of coding the sex of a respondent with the codes "1" and "2", should the coder attempt to enter any value other than "1 or 2", the program would automatically reject the entry and ask that a correct value be entered. However, this type of program still cannot handle cases where "1" is the correct response yet the coder mistakenly enters "2"—a permissible but incorrect code. Data-entry programs can also be tailored so that if respondents were instructed to skip questions, the program would automatically bypass data entry for these skipped questions (automatically entering numeric codes for "not applicable").

DESCRIBING AND EXPLAINING RESULTS

Frequency Tables

In reporting survey results, basic descriptive information about each question is typically generated. Occasionally, the only conclusions in which a surveyor is interested are based on this descriptive information. For example, a political campaign manager might wish to know how many people favour her or his candidate, while an employment consultant might wish to know how many firms have parental leave policies, or how many job vacancies are advertised by word-of-mouth.

Answers to all of these questions require single values, which may be expressed either as a raw number or as a percentage. However, the information that 715 people favour Candidate A is not particularly useful because no context is provided (i.e., how many people were asked). By adding the fact that 1200 people answered the question, the number 715 is placed in perspective and a *relative frequency* is expressed. Now a quick calculation reveals to us that, in relation to the total number of people canvassed, a majority of people favour Candidate A. This majority can be expressed even more precisely if we describe the number of people favouring Candidate A as a percentage of all valid answers. Of those providing a usable answer, 59.6 percent favour Candidate A ($[715 \div 1200] \times 100$). By expressing the frequency as a percentage, we are effectively saying that for every 100 people, 59.6 support Candidate A. When expressed as a percentage, therefore, the information is standardized, making comparison much easier. It is always wise to avoid the temptation of carrying results to several decimal places (e.g., 59.5833 percent) because it gives an air of false precision. Furthermore, if the number of cases on which a percentage is based is low (i.e., less than 20), do not calculate percentages.

In presenting the data in this manner, it is also important to recall the sampling principles of Chapter 11. A statement needs to be made about both the margin of error and the confidence level. In this particular case, we would say that with 95 percent confidence, the percentage of people in the population that would vote for Candidate A lies between 56.8 and 62.4 percent (the margin of error being +/− 2.8 percent).

Another way to present the results of a political poll is to show the percentage calculations for all candidates. This information can be displayed in a *frequency table* (see Figure 13.8), which is regularly generated in most surveys. This table

FIGURE 13.8 VOTING PREFERENCES OF A SAMPLE: FREQUENCY TABLE

Value	Value Label	Frequency	Total %	Valid %
1	Candidate A	715	54.2	59.6
2	Candidate B	308	23.4	25.7
3	Candidate C	177	13.4	14.8
9	Missing Data	118	9.0	
	Valid Cases 1200	Missing Cases 118		

FIGURE 13.9 VOTING PREFERENCES OF A SAMPLE: PIE CHART

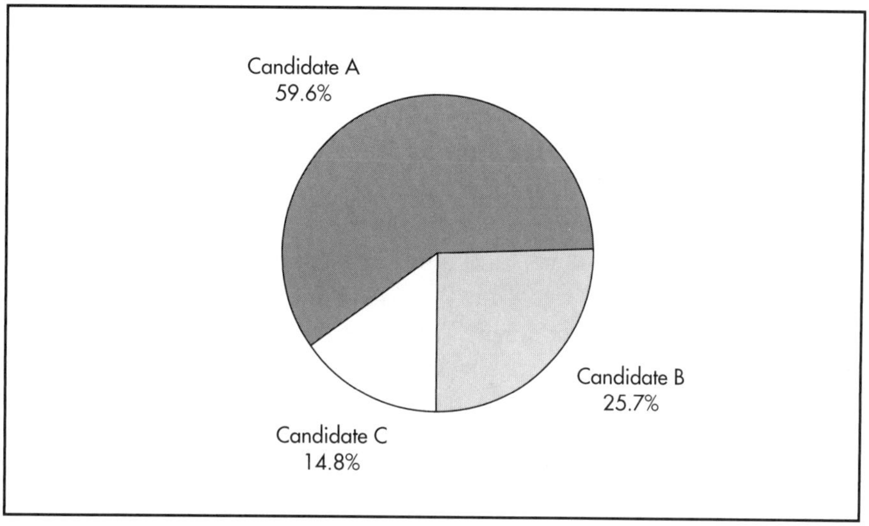

tells us that 1318 people were asked about the candidates, and of these, 1200 gave valid answers while 118 did not; that is, 9 percent of the respondents did not give a usable reply. The table also tells us the percentage of people favouring each candidate. This is most accurately seen in the "Valid %" column, where the proportion of people favouring each candidate is expressed as a percentage of all the valid, usable answers; in other words, the percentage values in this column are calculated only after the missing cases have been excluded.

In presenting tables of this kind, remember to provide an informative title. As well, be sure to include the number of cases on which the figures are based, along with a statement about missing data. The basic information contained in a frequency table can also be presented using graphic methods. A properly drawn pie chart is an effective way of presenting basic descriptive information (see Figure 13.9). Note, however, that the pie chart contains less detail than the frequency table, omitting, as is typical practice, the sample size and number of missing cases. Therefore, what a pie chart gains in visual simplicity often comes at the expense of information provided (relative to the frequency table). These advantages and disadvantages must be weighed before selecting a method of presenting information.

Pie charts or frequency tables work best when there is a relatively short list of values for any one variable. However, when dealing with questions such as age or income, which contain many possible values, the researcher must compute descriptive summarizing statistics, like the average or the median, in order to provide basic information.

Contingency Tables

There are occasions when our primary interest is in examining the relationships between variables—we want to answer questions about *why* certain patterns emerged. For example, based on our frequency table results, someone might be interested in why Candidate B or C received fewer votes than A. Was there, in fact, any particular pattern to be found in the type of people favouring each candidate? Sex of the respondent might be one variable influencing candidate choice. Figure 13.10 illustrates how we might check to see whether candidate preference differs between women and men. This type of table is referred to as either a cross-tabulation (a "cross tab") or a *contingency table,* which comes from the idea that candidate preference might be contingent on the sex of the respondent.

Before examining the numbers in the table, note how the table is arranged. A short title is used to describe the two variables that are being related. The labels of the independent variable ("Sex") and the dependent variable ("Voting Preference") are clearly specified, and the value labels for the categories of each variable are displayed. If either the variable or value labels require greater clarification, details should be provided in a footnote accompanying the table.

In laying out a contingency table, place the independent variable across the top of the table, the values of the dependent variable along the side, and then calculate the percentages so that the columns each add up to close to 100 percent; that is, calculate relative frequencies of the dependent variable (Candidate A, B, or C) for each separate category of the independent variable (sex of the respondent). In the table shown in Figure 13.10, the first column of percentages shows the relative frequency of women choosing Candidate A, B, or C, and

FIGURE 13.10 VOTING PREFERENCE BY SEX: CONTINGENCY TABLE

	Sex	
Voting Preference	**Women**	**Men**
Candidate A	40.2%	80.3%
Candidate B	44.5%	5.5%
Candidate C	15.3%	14.1%
Column Totals*	100.0%	99.9%
Number of Cases	620	580

Respondents were asked, "Which candidate would you vote for if an election were held today?"

*Note that total may not equal 100 due to rounding.

the column of percentages for women totals 100 percent. Likewise for male respondents, the percentage choosing each candidate is calculated. Notice that in our example the total percentage number for male respondents is 99.9, not 100, as a result of rounding; a note is included in the table to explain this. Finally, beneath the column percentages the number of women and men is reported. This latter information is crucial so that readers will know how many cases were included in the percentage calculations. It is important, too, that these raw numbers are displayed in such a way that the exact number of women and men choosing each candidate can be reconstructed by readers of the table.

To answer the initial question of whether there is a sex difference in voting preference, simply compare the percentages horizontally (i.e., across a row). The table shows us that 40.2 percent of women chose Candidate A, as compared to 80.3 percent of men, and that Candidate B received 44.5 percent of the female vote but only 5.5 percent of the male vote. Percentage differences, calculated within rows (e.g., 80.3 − 40.2 = 40.1), show large discrepancies between the voting preferences of women and men.

As long as the table is arranged with the independent variable across the top, the following rule is a useful way of analyzing tables: *percentage down, compare across*. With percentages running down the table and totalling 100 percent at the bottom, read across the table to determine percentage differences. This general rule takes advantage of the causal logic discussed in Chapter 5. Survey researchers are often interested in the generic question "What difference does a difference make?"; or, phrased more precisely, "What difference does the independent variable make to the distribution of the dependent variable?"; or, considered with respect to the above example, "What difference does sex make to the distribution of voting preferences?" Tables can be constructed and percentaged incorrectly (i.e., in the wrong direction), so be careful in the layout and calculations.

Figure 13.10 contains two variables and can therefore be referred to as a *bivariate* contingency table. To test the ideas of spuriousness discussed in Chapter 5, the researcher would have to add a third variable. Men and women differ systematically across many dimensions but for illustrative purposes we have selected labour force participation. Conceivably, both sex and voting preference might be related to labour force involvement, and it would be wise to check that the large sex differences detected in Figure 13.10 are not a function of women and men having different connections to the labour force (where, for example, more women hold part-time positions). A relatively easy way to examine this possible spurious factor would be to divide the sample on the basis of labour force involvement (e.g., create two sub-samples—one based on labour force participants and the other on non-labour force participants). The basic table shown in Figure 13.10 could then be reconstructed for each separate sub-sample and the percentage differences recalculated. If the same percentage differences that appeared in the original Figure 13.10 hold in each of the new sub-tables, the original relationship between sex and voting preference was not spurious. Alternatively, the newly calculated percentage differences could differ in a variety of ways that would lead to new conclusions about the relationship between sex, labour force status, and voting preference (recall from Chapter 4 that labour force status would be an intervening variable). With the introduction of third variables, the resulting interpretations of *multivariate* contingency tables grow increasingly complex and well beyond the scope of this book.

Table Etiquette

The following table etiquette should be followed when constructing contingency tables:

1. Use short, clear titles naming the key variables in the table.
2. Use labels for variables and their values that are meaningful to readers.
3. Always report the number of cases on which percentages are based.
4. Clearly show the direction in which the percentages total 100.
5. Present clarifying footnotes that give readers necessary definitional material.

The basic features of both frequency and contingency tables can also be displayed in graphic form. Just as a pie chart provides a visual display of frequency data, a bar chart provides an equivalent medium for tabular data. The basic percentages displayed in Figure 13.10 are represented as vertical bars in Figure 13.11, where the heights of the bars reflect differences in the voting preferences of women and men. As with pie charts, informational detail is lost in bar charts relative to contingency tables.

WRITING A FINAL REPORT

The last step in the survey research process—writing a report—demands as much attention to detail and quality as did the previous steps. All the painstaking work involved in carrying out a survey is rendered ineffective if the report does not hold the reader's attention and convey significant information in a meaningful fashion.

Most survey research reports share the same basic organization. The following outline can serve as a guide to the elements in a final report and their order of presentation:

1. a summary, an executive summary, or an abstract;
2. a brief discussion of the research problem or issue;

FIGURE 13.11 VOTING PREFERENCE BY SEX: BAR CHART

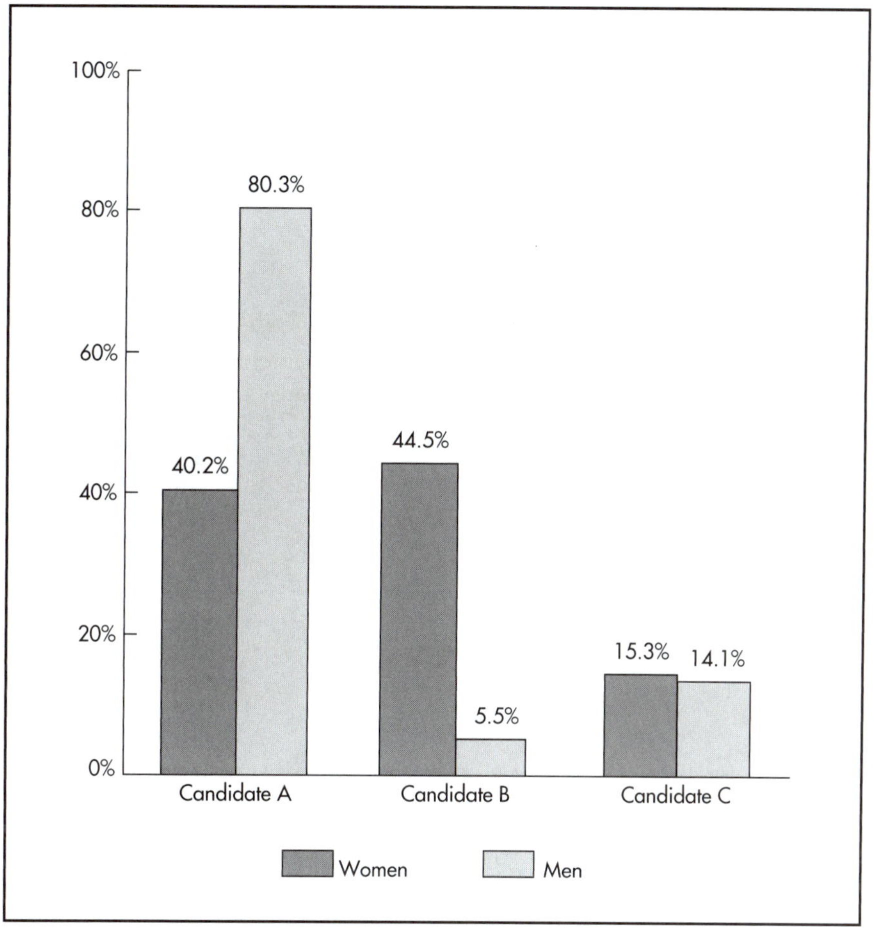

3. a brief discussion of the ideas and/or literature related to the research question;
4. the presentation of the research methodology;
5. the disposition of the sample;
6. the presentation of the findings; and
7. the implications, recommendations, and conclusions.

The Summary or Abstract

The summary is also referred to as the *executive summary* or the *abstract*. The summary allows readers to: (1) see what the report contains without having to read every page, (2) review the report before reading the complete text,

(3) understand the structure of the report, and (4) focus on the key points of the report.

The executive summary presents the key ideas in point form; it is like an outline, but the points are usually made as statements or sentences (see Figure 13.12). Executive summaries can extend over two or three pages, but most are contained on one page, allowing the reader to see key points at a glance. Reports over 10 pages should also include a table of contents, with page numbers locating each section.

The Problem

This section of the report is a summary of the problem development described in Chapter 4. It starts with a statement of the research question, and then describes the context in which the research problem is cited, including why the solution to the problem is important.

FIGURE 13.12 SAMPLE EXECUTIVE SUMMARY

Executive Summary

The Problem:

- The department heads at Coast General Hospital are not working together as a team.

The Method of Analysis:

- Interviews were conducted with managers and with staff from all departments.

The Results:

- The problems in team building at Coast General Hospital stem from
    - poor communication between some long- and short-term managers,
    - resistance to change from some long-term managers,
    - lack of training of managers in management skills,
    - too many managers in the management team, and
    - the low level of competence of some managers.
- The managers of Logistical Services and Administration are performing at a low level of management competence. Staff morale is low. Little direction is provided to staff. Poor communications are the norm rather than the exception.

The Recommendations:

- The services of the managers of Logistical Services and Administration be terminated.
- On-site management training for managers be implemented.
- The organizational structure be modified to reduce the number of managers (see Charts 2–3).

The Reasoning

This section shows the development of the theory and/or hypotheses that were tested. An example would be the material developed for Chapter 4. Theoretical and research literature that relates to the problem is summarized and evaluated. The reasoning behind the theory is developed. Hypotheses that follow from the reasoning are stated.

The Method of Analysis

This section describes the questionnaire that was used, particularly how specific questions provided measures of the key variables. It also describes the administration of the survey (including a discussion of the type and timing of data-collection procedures), and provides details on the sample-selection procedures, the size of the sample, and the statistical measures to be used in analyzing the results.

Disposition of the Sample The disposition of the sample—that is, the number and percentage of people/units who responded, as well as those who refused to participate or did not otherwise respond—is important when establishing generalizations that apply to the population. The researcher's ability to generalize the sample findings is related to the representativeness of the sample. Generalizations can be especially hazardous if respondents who completed the survey are different from those who did not. As we noted in Chapter 12, one way to check for representativeness is to compare the socio-demographic characteristics of the sample to the population; such details are reported in this section.

The Results

The presentation of results usually starts with tables that describe each of the key variables, followed by the testing of the theory or hypotheses. The important part of the presentation of results is to take the reader through the thinking processes that led to the development of the theory. If the intent of the research was to learn something about what influences voter decisions, product purchases, or employee behaviour, that issue must remain as the central focus. The central theme must flow smoothly through the entire results section. In particular, avoid becoming caught up in exploring tangents that are not central to the survey purpose.

Within the central theme, the reader should be guided step by step through the reasoning process. Determine the point that you are trying to make by using and referring to tables, charts, or graphs. If you are going to use a table, keep it simple. Each table should present one finding—the point you are trying to make. If a table has more than one result, it is better to break it up into two or more smaller, simpler tables. Use a chart or graph if it can illustrate the point more effectively; in this case, the tables can be placed in an appendix. The end of this section functions in much the same way as it does for any general essay or term paper. The findings section focuses on specific tables, graphs, and/or charts, and the end of this section ties together all the findings. Has the theory or idea been confirmed, and why is this conclusion significant?

Recommendations

Based on the research findings, this section is used to describe what course or courses of action are being recommended. In other words, what has been learned, and what are the implications for action, policies, or programs? This section should also provide the reader with a clear sense of just how solid the recommendations are. Qualifications should also be noted where appropriate.

SUMMARY

This chapter has focused on the processes involved in translating respondents' answers into a computerized form. A set of explicit rules for translation must be developed, with a code assigned for every survey answer and any single answer coded with one—and only one—value. These rules are summarized in a codebook. Another set of rules applies to the exact location of data in a computer file. These rules allow computer programs to read the data files and to summarize or analyze the information of all respondents. Finally, the summarized data can be displayed in the form of frequency tables or pie charts, or, alternatively, contingency tables or bar charts.

The research process is complicated and demanding of time and energy. Somewhat like running a race or coming to the final scene in a play, you need to keep momentum and energy through to the finish. In this case, writing a final report is your presentation of all the prior work. In an important sense, your careful work is only as good as your final report, which should highlight the high quality and professionalism of all your survey research decisions and practices.

EXERCISE 13.1 CODEBOOK CONSTRUCTION

1. Develop a codebook for transferring information from your questionnaires to the computer. Use Exercise 6.1 as a starting point.

EXERCISE 13.2 CODING AND ERROR-CHECKING INSTRUCTIONS

1. Develop a set of instructions showing coders how to use the codebook in transferring data from your questionnaire to the computer-readable form.

2. Develop a set of procedures for determining if coding errors are being made. Establish a quality control standard. Develop a set of procedures that are invoked when the quality control standards are not being met.

EXERCISE 13.3 COMPLETED REPORT

Produce a final report. The report should have a cover and a table of contents; a letter of transmittal to the research sponsor; a statement of the research question and the development of the theory; a statement of hypothesis and the types of questions; a description of methodology, including the sample frame; information on the administration of the survey (e.g., mail-out waves, response rates, Web design); and some possible sample findings, with mock tables, summary, and conclusions. Use mock data. Provide a copy of the questionnaire, codebook, administrative materials, and so forth in appendices.

FURTHER READING

Bainbridge, W.S. (1989). *Survey research: A computer-assisted introduction.* Belmont, CA: Wadsworth Publishing.

Buckley, J. (2004). *Fit to print: The Canadian student's guide to essay writing* (6th ed.). Toronto, ON: Thomson Nelson.

Knoke, D., Bohrnstedt, G.W., and Mee, A.P. (2002). *Statistics for social data analysis* (4th ed.). Belmont, CA: Wadsworth Publishing.

Marsh, C. (1988). *Exploring data.* Cambridge: Polity Press.

Saris, W.E. (1991). *Computer-assisted interviewing.* Newbury Park, CA: Sage.

Appendix

This appendix includes excerpts from several different questionnaires. The examples have been reformatted to illustrate visual cues often used in questionnaires. To illustrate several different techniques, some visuals used in this appendix are not consistent with one another. However, remember that actual questionnaires must use consistent formatting.

Section A is revised from Statistics Canada's *General Social Survey: Where Does Time Go?* (1991). It illustrates questions about the demographic attributes of respondents.

SECTION A **DEMOGRAPHICS** This section asks questions about your demographic attributes. It is helpful in making comparisons between categories, such as geographic regions or educational levels.

A1. In what country were you born?

- ❏ Canada → In which province or territory?
  - ❏ Newfoundland
  - ❏ Prince Edward Island
  - ❏ Nova Scotia
  - ❏ New Brunswick
  - ❏ Quebec
  - ❏ Ontario
  - ❏ Manitoba
  - ❏ Saskatchewan
  - ❏ Alberta
  - ❏ British Columbia
  - ❏ Yukon Territory
  - ❏ Northwest Territories
- ❏ Country outside of Canada *(specify)* ____________

A2. What is your date of birth?

|_|_| Day |_|_| Month |_|_|_|_| Year

A3. To which ethnic or cultural group do you or did your ancestors belong? *(Accept multiple response)*

- ❏ African
- ❏ Arabic
- ❏ Chinese
- ❏ East Indian
- ❏ English
- ❏ First Nations
- ❏ French
- ❏ German

(continued)

(Section A, continued)

- ❑ Irish
- ❑ Italian
- ❑ Japanese
- ❑ Korean
- ❑ Scottish
- ❑ Ukrainian
- ❑ Other *(specify)* ____________

A4. What, if any, is your religion?

- ❑ No religion
- ❑ Anglican
- ❑ Buddhist
- ❑ Hindu
- ❑ Islamic
- ❑ Jewish
- ❑ Protestant
- ❑ Roman Catholic
- ❑ Other *(specify)* ____________________________

A5. Are you:

- ❑ Male
- ❑ Female

A6. What was your income before taxes, from wages, salaries, and self-employment during the last 12 months?

- ❑ Income ⟶ $ _________
- ❑ Loss ⟶ $ _________
- ❑ No Income
- ❑ Don't know

A7. What was your income during the last 12 months from government sources such as Family Allowance, E.I., Social Assistance, Canada or Quebec Pension Plan or Old Age Security?

$ ___________

- ❑ No Income
- ❑ Don't know

A8. What is your best estimate of the total income of all household members from all sources during the last 12 months? Was the total household income… [Note: This is typically used in telephone interviews]

- ❑ Less than $20,000 ⟶
  - ❑ Less than $10,000 ⟶
    - ❑ Less than $5,000
    - ❑ $5,000 and more
  - ❑ $10,000 and more ⟶
    - ❑ Less than $15,000
    - ❑ $15,000 and more

(continued)

(Section A, continued)

- ❑ $20,000 and more →
  - ❑ Less than $40,000 →
    - ❑ Less than $30,000
    - ❑ $30,000 and more
  - ❑ $40,000 and more →
    - ❑ Less than $60,000
    - ❑ $60,000 and more

❑ No Income

❑ Don't Know

A9. How many years of elementary and secondary education have you completed?

❑ No schooling
❑ One to five years
❑ Six
❑ Seven
❑ Eight
❑ Nine
❑ Ten
❑ Eleven
❑ Twelve
❑ Thirteen

A10. Have you graduated from secondary school?

❑ Yes
❑ No

A11. Have you had any further schooling beyond elementary/secondary school?

❑ Yes
❑ No ——→ *(go to A14)*

A12. What was the highest level you attained?

❑ Some community college, CEGEP, or nursing school
❑ Diploma or certificate from community college, CEGEP, or nursing school
❑ Some university
❑ Bachelor or undergraduate degree or teacher's college
❑ Master's or earned doctorate
❑ Other *(specify)*

____

❑ Not applicable: I am currently a student

(continued)

(Section A, continued)

A13. In which year did you reach your highest level of education?

|_|_|_|_|

❑ Year ❑ Not applicable: I am currently a student

A14. What best describes your main activity for the last four weeks?

❑ Working for pay → *(go to A15)*
❑ Student → *(go to A18)*
❑ Unemployed → *(go to A18)*
❑ Retired → *(go to A18)*
❑ Other *(specify)*
____________________ *(go to A18)*

A15. Were you an employee working for someone else or self-employed?

❑ An employee working for someone else
❑ Self-employed/employer

A16. What was the main kind of business, industry, or service? *(Give a full description: e.g., paper box manufacturing, retail shoe store, municipal board of education)*

____
____
____

A17. What kind of work were you doing in this position? *(Give a full description: e.g., posting invoices, selling shoes, teaching primary school)*

____
____
____

A18. Are you:

❑ Married
❑ Separated
❑ Divorced
❑ Common-law
❑ Single
❑ Widowed

Source: Adapted from Statistics Canada (1991), *General social survey: Where does time go?* Cat. no. 8-4500-33-1.

Researchers sometimes ask questions using vignettes, where the respondent is given a descriptive scene, then asked to respond to questions about, or based on, the vignette. The following questions about sexual aggression are drawn from a study by Hutchinson (1994), measuring students' perception of the degree of aggression involved in various vignettes.

SECTION B

PERCEPTIONS OF AGGRESSION This section asks questions about sexual aggression. A vignette is presented, followed by a scale ranging from "No Sexual Aggression" to "Severe Sexual Aggression." The abbreviation "SA" refers to "sexual aggression." *In questions B1 and B2, please circle the appropriate number corresponding to your perception of the vignette situation.*

B1. Kathy has invited a friend and her boyfriend over to her place to play Scruples. Since they need a fourth player in order to play the game, Kathy leaves it up to her friend to invite someone along. Kathy's friend, Laurie, brings along one of her boyfriend's friends (Nick). The evening goes well, with Kathy and Nick getting to know each other. While Kathy is in the kitchen getting snacks for her guests, Nick comes in to help her. He comes up behind her and reaches around to grab her breasts. Kathy turns around and tells Nick, "Let's just go back into the other room and forget this ever happened." Nick just stares at her.

1	2	3	4	5
No Sexual Aggression	Mild SA	Moderate SA	High SA	Severe Sexual Aggression

B2. A group of people who work together in a large organization decide to have a Super Bowl party. Christie doesn't drink alcohol and brings her own club soda. Anthony introduces himself to Christie and asks to share some of her soda since he doesn't drink either. Everyone happens to be a San Francisco 49ers fan. As the time expires and their team wins, everyone hugs each other and celebrates. As Christie and Anthony hug, he puts his hand between her breasts. This offends Christie, and she pulls away from Anthony. Anthony immediately walks away.

1	2	3	4	5
No Sexual Aggression	Mild SA	Moderate SA	High SA	Severe Sexual Aggression

Source: R. Hutchinson (1994), Students' perception of male sexually aggressive behaviour, *Sex Roles, 30*, 407–22.

Surveys are also conducted on units other than individuals, such as organizations, hospitals, schools, or companies. Statistics Canada conducts and publishes an annual survey called *Manufacturing Industries of Canada*. The following is revised from the 1995 edition, to illustrate a study done on organizations. Notice the code numbers beside all response categories. See Chapter 13 of this book for coding details.

SECTION C

ORGANIZATIONS This section is concerned with information about manufacturing businesses in Canada. These questions are concerned with the manufacturing organization that you represent for this reporting year.

C1. Did this establishment go out of business during the reporting year?

1 ❑ Yes Date__________
2 ❑ No

C2. Did any change of ownership occur during the reporting year?

1 ❑ Yes Date__________
2 ❑ No

If "Yes," provide information for the full reporting year. If this is not possible, report for the period operated and give name, address and telephone number of person to contact for the balance of the data.

Name__________ Phone number __________
Address __________

C3. Type of organization *(check one)*

1 ❑ Individual ownership *(skip to C5)*
2 ❑ Partnership *(skip to C5)*
3 ❑ Co-operative *(skip to C5)*

4 ❑ Incorporated company

C4. If your company is incorporated, please provide your Corporation Account Number (8 digits) or, if your Corporation Account Number has been converted to the new Business Number (9 digits followed by RC), please provide your Business Number. You may obtain these numbers from your latest Canada Customs and Revenue Agency Assessment Notice (form T456) or your Canada Customs and Revenue Agency Taxation Remittance Form (T9RC).

Corporation Account Number ☐☐☐☐☐☐☐☐ OR Business Number ☐☐☐☐☐☐☐☐☐

Now to C5

(continued)

(Section C, continued)

C5. Does this establishment have a Canadian Head or Executive Office whose operations can be reported separately?

1 ❑ Yes
2 ❑ No

Source: Adapted from Statistics Canada (1995), *Manufacturing industries of Canada: National and provincial areas,* Cat. no. 31-203-XPB.

Researchers sometimes ask questions that respondents may find psychologically threatening. Threatening or sensitive questions can cause the respondent distress if the question reminds the respondent of a troubling incident or probes into a sensitive topic. The questions listed below illustrate questions about a personal and sensitive topic regarding health, sexuality, and control over bodily functions. This is revised from a study by Fowler et al. (1998) on men who have had prostate surgery.

SECTION D **HEALTH** This section contains questions about health and sexuality. It may be distressing for some respondents; however, they are very important questions. *Please check the appropriate responses.*

D1. Do you still have any problem at all with dripping or leaking urine?

01 Yes ....❑
02 No ....❑

D2. In the *past month*, have you dripped or leaked urine when you coughed or sneezed?

01 Yes ....❑
02 No ....❑

(continued)

(Section D, continued)

D3. Some men wear pads, rubber pants, adult diapers, or a clamp to help with wetness. Do you use anything like that now?

01 No ....❑ *(skip to D5)*

02 Yes ....❑

D4. What kind of wetness control do you use?

01 Pads ....❑
02 Rubber pants ....❑
03 Adult diapers ....❑
04 Clamps ....❑
05 Other *(specify)* ....❑

D5. Since your prostate surgery, have you had any full erections?

01 Yes ....❑
02 No ....❑

D6. Have you had any partial erections?

01 Yes ....❑
02 No ....❑

D7. Over the past month or so, if you have had erections, how often were they firm enough to have intercourse?

01 Never ....❑
02 A few times ....❑
03 Fairly often ....❑
04 Usually ....❑
05 Always ....❑

Source: F.J. Fowler, A.M. Roman, and Zhu Xiao Di. (1998), Mode effects in a survey of medicare prostrate surgery patients, *Public Opinion Quarterly, 62*, 29–46.

Often, researchers will want to ask respondents about some repeated phenomenon, and often these questions will cover a span of time. This section, modified from Fiona Kay's study of women in law, entitled *Transitions in the Ontario Legal Profession* (1991), asks a repeated set of questions about the various positions the respondent has held in the legal profession since her first position.

SECTION E

FIRST POSITION HELD AFTER BEING CALLED TO THE BAR

This section contains questions about the first position you held after being called to the Bar. It allows comparisons to be made between different positions, and over time.

E. Is your current position the first position you have held since being called to the Bar? *(Check one)*

❑ Yes → *(Go to Section II: "Current Position")*
❑ No *(Go to E1)*

Thinking about your first professional position after being called to the Bar

E1. When did this position start and end?

Start date: |_|_| Month |_|_|_|_| Year End date: |_|_| Month |_|_|_|_| Year

E2. What type of position was this? *(Check one)*

❑ Full-time ❑ Part-time Other *(Specify)* ____________________

E3. Which of the following describe this postion? *(Check as many as apply)*

Were you:

- ❑ not engaged in the practice of law *(Go to E10)*
- ❑ employee of government
- ❑ employee of corporation
- ❑ employed with Legal Aid or law clinic
- ❑ employed in private industry
- ❑ employee or associate of law firm
- ❑ sole practitioner
- ❑ partner
- ❑ Other *(Specify)* ____________________

E4. Including yourself, how many lawyers worked in your office (that is, at one physical location) at the time you left for your next position?

________________ lawyers

E5. What were the main fields of law you practised during your time in this position? *(Check as many as apply):*

- ❑ Social welfare
- ❑ Other administrative
- ❑ Corporate commercial
- ❑ Banking
- ❑ Industrial and intellectual property
- ❑ Labour
- ❑ Real estate
- ❑ Tax

(continued)

(Section E, continued)

- ❑ Securities
- ❑ Criminal
- ❑ Wills and estates
- ❑ Family
- ❑ Civil litigation
- ❑ Constitutional
- ❑ Other *(Please specify)* ____________________

E6. What was your approximate gross annual income when leaving this position? (*after business deductions and before taxes*)

$ __________

E7. Was there was any time interval between leaving this position and starting your next position?

- ❑ No *(Skip to E10)*
- ❑ Yes

E8. Please specify the number of months in this time interval.
____________________ months

E9. What was your primary activity during this period?

- ❑ travel
- ❑ further education
- ❑ looking after my children
- ❑ unemployed
- ❑ illness or injury
- ❑ other *(Specify)* ____________________

E10. Is the next position you held your current position?

- ❑ Yes ⟶ *(Go to Section II: "Current Position")*
- ❑ No ⟶ *(Go to Section I: Part F: "Second Position Held After Being Called to the Bar")*

Source: Adapted from F. Kay, *Transition in the Ontario legal profession: A survey of lawyers called to the Bar between 1975 and 1990*, The Law Society of Upper Canada.

The respondents would complete sections provided until they answered "yes" to the question "Is the next position you held your current position?" The researcher needs to know the maximum number of positions a person might have held to ensure that enough sections exist to accommodate them all.

References

Alasuutari, P. (1995). Beyond the qualitative quantitative distinction: Crosstabulation in qualitative research. *International Journal of Contemporary Sociology*, 32(2), 251–68.

Ames, H.B. (1897) [1972]. *A city below the hill: A sociological study of a portion of the city of Montreal.* Montreal, QC: Bishop Engraving and Printing (1897); Toronto, ON: University of Toronto Press (1972).

Babbie, E. (1990). *Survey research methods* (2nd ed.). Belmont, CA: Wadsworth.

Bachman, J.G., and O'Malley, P.M. (1981). When four months equals a year: Inconsistencies in student reports of drug use. *Public Opinion Quarterly, 45*(4), 536–48.

Bainbridge, W.S. (1989). *Survey research: A computer-assisted introduction.* Belmont, CA: Wadsworth.

Baker, R. (1992). New technology in survey research: Computer-assisted personal interview. *Social Science Computer Review, 10*(2), 145–57.

Barnes, J. (1979). *Who should know what? Social science, privacy and ethics.* New York: Penguin.

Barnette, V. (1991). *Sample survey principles and methods.* London: Edward Arnold.

Bearden, W.O., Netemery, R.G., and Mobley, M.F. (1993). *Handbook of marketing scales: Multi-item measures for marketing and consumer behavior research.* Newbury Park, CA: Sage.

Becker, H.S. (1998). *Tricks of the trade: How to think about your research while you're doing it.* Chicago: University of Chicago Press.

Berk, R.A., and Rossi, P.H. (1990). *Thinking about program evaluation.* Newbury Park, CA: Sage.

Bishop, G.F., et al. (1980). Pseudo-opinions on public affairs. *Public Opinion Quarterly, 44*(2), 198–209.

Blake, D.E. (1985). *Two political worlds: Parties and voting in British Columbia.* Vancouver, BC: University of British Columbia Press.

Bohrnstedt, G., and Knoke, D. (1988). *Statistics for social data analysis* (2nd ed.). Itasca, IL: Peacock.

Booth, C. (Ed.). (1892). *Life and labour of the people of London* (Vol. 1). London: Macmillan.

Bositis, D.A. (1990). *Research designs for political science: Contrivance and demonstration in theory and practice.* Carbondale, IL: Southern Illinois University Press.

Bottomore, T., and Rubel, M. (Eds.). (1956). *Karl Marx: Selected writings in sociology and social philosophy.* New York: McGraw-Hill.

Bradburn, N., and Sudman, S. (1979). *Improving interview method and questionnaire design.* San Francisco, CA: Jossey-Bass.

Brick, M.J., et al. (1995). Bias in list-assisted telephone samples. *Public Opinion Quarterly, 59*, 218–35.

Brin, R., and Hague, P.N. (2000). *The handbook of international market research techniques*. London: Kogan Page.

Buckley, J. (2001). *Fit to print: The Canadian student's guide to essay writing* (5th ed.). Toronto, ON: Nelson Thomson Learning.

Bulmer, M., Bales, K., and Sklar, K.K. (1991). *The social survey in historical perspective, 1880–1904.* New York: Cambridge University Press.

Campbell, D.T., and Stanley, J.C. (1969). *Experimental and quasi-experimental designs for research.* Chicago, IL: Rand McNally.

Cannell, C.F., Marquis, K.H., and Laurent, A. (1977). A summary of studies of interviewing methodologies. *Vital and Health Statistics.* Washington, DC: U.S. Public Health Service.

Catania, J., et al. (1996). Effects of interviewer gender, interviewing choice and item wording on responses to questions concerning sexual behaviour. *Public Opinion Quarterly, 60*(3), 345–75.

Census of the Canadas. (1854). *Second report of the secretary of the board of registration and statistics, on the census of the Canadas for 1851–52.* Quebec City, QC: John Lovell.

Cha, Y.S. (2005). All that glitters is not gold: Examining the perils and obstacles in collecting data on the internet. *International Negotiation, 10*(1), 115–30.

Choinière, M., and Amsel, R. (1996). A visual analogue thermometer for measuring pain intensity. *Journal of Pain and Symptom Management, 11*(5), 299–311.

Cornell, L.L., and Hayami, A. (1986). The Shumon Aratame Cho: Japan's population registers. *Journal of Family History, 11*(4), 311–28.

Couper, M.P. (2000). Web surveys: A review of issues and approaches. *Public Opinion Quarterly, 64*(4), 464–94.

Couper, M.P., and Burt, G. (1994). Interviewer attitudes toward CAPI. *Social Science Computer Review, 12*(1), 38–54.

Couper, M.P., Tourangeau, R., Conrad, F., and Crawford, S. (2004). What they see is what they get: Response options for Web surveys. *Social Science Computer Review, 22*(1), 111–27.

Couper, M.P., Tourangeau, R., and Kenyon, K. (2004). Picture this! Exploring visual effects in Web surveys. *Public Opinion Quarterly, 68*(2), 255–66.

Couper, M.P., Traugott, M., and Lamias, M. (2001). Web survey design and administration. *Public Opinion Quarterly, 65*(2), 230–53.

Curtin, R., Presser, S., and Singer, E. (2005). Changes in telephone survey nonresponse over the past quarter century. *Public Opinion Quarterly, 69*(1), 87–98.

Czaja, R., and Blair, J. (1996). *Designing surveys: A guide to decisions and procedures.* Thousand Oaks, CA: Sage.

de Rada, V.D. (2005). Influence of questionnaire design on response to mail surveys. *International Journal of Social Research Methodology, 8*(1), 61–78.

Dillman, D. (1978). *Mail and telephone surveys: The total design method.* New York: John Wiley.

Dillman, D. (1991). The design and administration of mail surveys. *Annual Review of Sociology, 17*, 225–49.

Dillman, D. (2000). *Mail and Internet surveys: The tailored design method.* New York: John Wiley.

Dillman, D., and Bowker, D. (2001). The Web questionnaire challenge to survey methodologists. In U. Reips and M. Bosnjak (Eds.), *Dimensions of Internet science*. Lengerich, Germany: Pabst Science Publishers. Retrieved June 24, 2002, from http://survey.sesrc.wsu.edu/dillman/papers.htm.

Dillman, D., and Schaefer, D. (1998). Development of a standard e-mail methodology: Results of an experiment. *Public Opinion Quarterly, 62*, 378–97.

Dutka, S., and Frankel, L.R. (1993). Measurement error in organizational surveys. *American Behavioral Scientist, 36*(4), 472–84.

Eichler, M. (1991). *Nonsexist research methods: A practical guide.* New York: Routledge.

Floyd, R. (1995). *Success in the social sciences: Writing and research for the Canadian student*. Toronto, ON: Harcourt Brace.

Foddy, W. (1993). *Constructing questions for interviews and questionnaires.* Cambridge: Cambridge University Press.

Foschi, M. (1997). On scope conditions. *Small Group Research, 28*, 535–55.

Fowler, F., and Mangione, T. (1990). *Standardized survey interviewing: Minimizing interviewer-related error*. Newbury Park, CA: Sage.

Fowler, F.J. (2002). *Survey research methods* (3rd ed.). Newbury Park, CA: Sage.

Fowler, F.J., Roman, A.M., and Zhu Xiao Di. (1998). Mode effects in a survey of medicare prostate surgery patients. *Public Opinion Quarterly, 62*, 29–46.

Fox, J., Murray, C., and Warm, A. (2003). Conducting research using Web-based questionnaires: Practical, methodological, and ethical considerations. *International Journal of Social Research Methodology,* 6(2), 167–80.

Gaziano, C. (2005). Comparative analysis of within-household respondent selection techniques. *Public Opinion Quarterly, 69*(1), 124–57.

Gendall, P. (2005). "Can you judge a questionnaire by its cover? The effect of questionnaire cover design on mail survey response. *International Journal of Public Opinion Research, 17*(3), 346–61.

Genovese, B.J. (2004). Thinking inside the box: The art of telephone interviewing. *Field Methods, 16*(2), 215–26.

Goyder, J. (1987). *The silent minority: Nonrespondents on sample surveys.* Cambridge: Polity.

Groves, R. (1990). Theories and methods of telephone surveys. *Annual Review of Sociology, 16*, 221–40.

Groves, R., et al. (1988). *Telephone survey methodology.* New York: John Wiley.

Groves, R.M., Fowler, F., Couper, M., Lepkowski, J., Singer, E., and Tourangeau, R. (2004). *Survey methodology*. Hoboken, NJ: Wiley InterScience.

Hagan, J., and McCarthy, B. (1997). *Mean streets: Youth crime and homelessness.* Cambridge University Press.

Heerwegh, D. (2005). Effects of personal salutations in e-mail invitations to participate in a Web survey. *Public Opinion Quarterly, 69*(4), 588–98.

Heerwegh, D., and Loosveldt, G. (2002). An evaluation of the effect of response formats on data quality in Web surveys. *Social Science Computer Review, 20*(2), 471–84.

Henry, G.T. (1990). *Practical sampling.* Newbury Park, CA: Sage.

Homan, R. (1991). *The ethics of social research.* New York: Longman.

Hoover, K. (2001). *The elements of social scientific thinking* (7th ed.). New York: St. Martin's Press.

Huskisson, E.C. (1983). Visual analogue scales. In Ronald Melzack (Ed.), *Pain measurement and assessment*, 33–37. New York: Raven Press.

Hutchinson, R. (1994). Students' perception of male sexually aggressive behaviour. *Sex Roles, 30* (March), 407–22.

Hutton, P. (1988). *Survey research for managers.* Basingstoke: Macmillan.

Jackson, W. (1988). *Research methods: Rules for survey design and analysis.* Scarborough, ON: Prentice Hall.

Kaplowitz, M.D., Hadlock, T.D., and Levine, R. (2004). A comparison of Web and mail survey response rates. *Public Opinion Quarterly, 68*(1), 94–101.

Kay, F. (1991). *Transitions in the Ontario legal profession: A survey of lawyers called to the Bar between 1975 and 1990*. The Law Society of Upper Canada.

Keeter, S. (2006). The impact of cell phone noncoverage bias on polling in the 2004 presidential election. *Public Opinion Quarterly, 70*(1), 88–98.

Keeter, S., et al. (2000). Consequences of reducing nonresponse in a national telephone survey. *Public Opinion Quarterly, 64*(2), 125–48.

Kemph, B.T., and Kasser, T. (1996). Effects of sexual orientation of interviewer on expressed attitudes toward male homosexuality. *Journal of Social Psychology, 136*, 401–03.

Keppel, G., and Zedeck, S. (1989). *Data analysis for research designs: Analysis-of-variance and multiple regression/correlation approaches*. New York: W.H. Freeman.

Kwong, J.C., et al. (2002). Effects of rising tuition fees on medical school class composition and financial outlook. *Canadian Medical Association Journal, 166*(8), 1023–28.

Lavrakas, P.J. (1993). *Telephone survey methods: Sampling, selection, and supervision* (2nd ed.). Newbury Park, CA: Sage.

Likert, R. (1932). A technique for the measurement of attitudes. *Archives of Psychology, 140*.

Marsh, C. (1982). *The survey method.* London: George Allen & Unwin.

Marsh, C. (1988). *Exploring data.* Cambridge: Polity Press.

McDonald, S. (2002). Is Canada the wired, Internet country we're told it is? Retrieved May 8, 2002, from http://www.bcentral.ca/archive/technology/survey_2002.asp.

Melzack, R. (1983). The McGill Pain Questionnaire. In Ronald Melzack (Ed.), *Pain measurement and assessment*, 41–47. New York: Raven Press.

Miller, G.A. (1956). The magical number seven, plus or minus two. *Psychological Review, 63*, 81–97.

Misanchuk, E.R. (1992). *Preparing instructional text: Document design using desktop publishing.* Englewood Cliffs, NJ: Educational Technology Publications.

O'Guinn, T.C., and Shrum, L.J. (1997). The role of television in the construction of consumer reality. *Journal of Consumer Research, 23*, 278–94.

Orr, S.K. (2005). New technology and research: An analysis of Internet survey methodology in political science. *Political Science and Politics, 38*(2), 263–67.

Palys, T. (2006). *Research decisions: Quantitative and qualitative perspectives* (4th ed.). Toronto, ON: Thomson Nelson.

Philipson, T. (1997). Data markets and the production of surveys. *Review of Economic Studies, 64*, 47–72.

Pineo, P., and Porter, J. (1967). Occupational prestige in Canada. *Canadian Review of Sociology and Anthropology, 4*(1), 24–40.

Presser, S. (1994). Informed consent and confidentiality in survey research. *Public Opinion Quarterly, 58*(3), 446–59.

Presser, S., and Schuman, H. (1980). The measurement of a middle position in attitude surveys. *Public Opinion Quarterly, 44*(1), 70–85.

Raymond, A. (1984). The population of Aleppo in the sixteenth and seventeenth centuries. *International Journal of the Middle East Studies, 16* (November), 447–60.

Rea, L.M., and Parker, R. (1997). *Designing and conducting survey research* (2nd ed.). San Francisco, CA: Jossey-Bass.

Reese, S.D., et al. (1986). Ethnicity of interviewer effects among Mexican-Americans and Anglos. *Public Opinion Quarterly, 50*, 563–72.

Richins, M.L. (1997). Measuring emotions in the consumption experience. *Journal of Consumer Research, 24*(2), 127–46.

Rosenfeld, P., Edwards, J.E., and Thomas, M.D. (Eds.). (1993). *Improving organizational surveys.* Newbury Park, CA: Sage.

Rossi, P. (1988). On sociological data. In Neil Smelser (Ed.), *Handbook of sociology*, 131–54. Newbury Park, CA: Sage.

Rubenstein, S.M. (1995). *Surveying public opinion.* Belmont, CA: International Thomson Publishing.

Saris, W.E. (1991). *Computer-assisted interviewing.* Newbury Park, CA: Sage.

Satin, A., and Shastry, W. (1993). *Survey sampling: A non-mathematical guide.* Ottawa: Statistics Canada.

Schwarz, N., et al. (1991). Rating scales: Numeric values may change the meaning of scale labels. *Public Opinion Quarterly, 55*, 570–82.

Singer, E., and Presser, S. (Eds.). (1989). *Survey research methods: A reader*, sect. 3, 187–244. Chicago, IL: University of Chicago Press.

Smith, T. (1987). That which we call welfare by any other name would smell sweeter: An analysis of the impact of question wording on response patterns. *Public Opinion Quarterly, 51*(1), 75–83.

Smith, T. (1990). The first straw poll: A study of the origins of election polls. *Public Opinion Quarterly, 54*, 21–36.

Smyth, J.D., Dillman, D., Christian, L., and Stern, M. (2006). Comparing check-all and forced-choice question formats in Web surveys. *Public Opinion Quarterly, 70*(1), 66–77.

Spaeth, J.L., and O'Rourke, D.P. (1994). Designing and implementing the National Organization Study. *American Behavioral Scientist, 37*(7), 872–90.

Spector, P.E. (1981). *Research design.* Beverly Hills, CA: Sage.

Statistics Canada. (1991) *General social survey: Where does time go?* Cat. no. 8-4500-33.1.

Statistics Canada. (1995). *Manufacturing industries of Canada: National and provincial areas.* Cat. no. 31-203-XPB.

Statistics Canada. (2005). Internet service provider industry. *The Daily*, Thursday, December 15, 2005.

Stewart, B.M. (1913). *Report of a preliminary and general social survey of Fort William/Port Arthur.* Department of Temperance and Moral Reform, Methodist Church and Department of Social Services and Evangelism, Presbyterian Church.

Streiner, D., and Norman, G. (1989). *Health measurement scales: A practical guide to their development and use.* New York: Oxford Medical Publications.

Sudman, S., and Bradburn, N. (1982). *Asking questions: A practical guide to questionnaire construction.* San Francisco, CA: Jossey-Bass.

Tanur, J.M. (1992). *Questions about questions: Inquiries into the cognitive bases of surveys.* New York: Russell Sage.

Taylor, W. (1990). *Social science research: Theory and practice.* Scarborough, ON: Thomson Nelson.

Tourangeau, R., and Smith, T.W. (1996). Asking sensitive questions: The impact of data collection mode, question format and question context. *Public Opinion Quarterly, 60*(2), 275–304.

Tryfos, P. (1996). *Sampling methods for applied research.* New York: John Wiley.

Van Selm, M., and Jankowski, N.W. (2006). Conducting online surveys. *Quality and Quantity*, *40*(3), 435–56.

Warriner, K., et al. (1996). Charities, no, lotteries, no, cash, yes: Main effects of incentives in a Canadian incentives experiment. *Public Opinion Quarterly, 60*, 345–75.

Willis, G. (2004). *Cognitive interviewing: A tool for improving questionnaire design.* London: Sage.

Wilson, J. (1988). *Analyzing politics: An introduction to empirical methods.* Scarborough, ON: Prentice Hall.

Zhan, Lin, et al. (1998). Promoting health: Perspectives from ethnic elderly women. *Journal of Community Health Nursing, 15*(1), 31–44.

Index